SNAP SCIENCE

GET YOUR TEETH INTO IT!

TEACHING FRAMEWORK
Year 6

Series Editor: Jane Turner

Series Consultant: Derek Bell

Authors:

Chris Banbury, Nicola Beverley

James de Winter, Bryony Turford

William Collins' dream of knowledge for all began with the publication of his first book in 1819. A self-educated mill worker, he not only enriched millions of lives, but also founded a flourishing publishing house. Today, staying true to this spirit, Collins books are packed with inspiration, innovation and practical expertise. They place you at the centre of a world of possibility and give you exactly what you need to explore it.

Collins. Freedom to teach.

Published by Collins
An imprint of HarperCollins*Publishers*
77–85 Fulham Palace Road
Hammersmith
London
W6 8JB

Browse the complete Collins catalogue at
www.collins.co.uk

© HarperCollins*Publishers* Limited 2014

10 9 8 7 6 5 4 3 2 1

ISBN 978-0-00-755146-0

The authors assert their moral rights to be identified as the authors of this work

All rights reserved. No part of this publication may be reproduced, stored in a retrieval system, or transmitted in any form or by any means, electronic, mechanical, photocopying, recording or otherwise, without the prior written permission of the Publisher or a licence permitting restricted copying in the United Kingdom issued by the Copyright Licensing Agency Ltd., 90 Tottenham Court Road, London W1T 4LP.

British Library Cataloguing in Publication Data

A Catalogue record for this publication is available from the British Library

Authors: Chris Banbury, Nicola Beverley, James de Winter, Bryony Turford
Publishing Manager: Lizzie Catford
Managing Editor: Helen Terrington
Development Editor: Melanie Hoffman
Project Editors: Mike Appleton, Alison Walters
Production: Rebecca Evans
Editors: Soo Hamilton, Sophia Ktori, Joan Miller, Christine Moorcroft, Sarah Vittachi
Cover design and artwork: Amparo Barrera
Internal design concept: Amparo Barrera
Designer: Ken Vail
Illustrations: Aptara Inc., Clive Goodyer, Ian Moores
Printed and bound by Martins the Printers Ltd, Berwick-upon-Tweed

FSC is a non-profi t international organisation established to promote the responsible management of the world's forests. Products carrying the FSC label are independently certifi ed to assure consumers that they come from forests that are managed to meet the social, economic and ecological needs of present and future generations, and other controlled sources.

MIX
Paper from
responsible sources
FSC™ C007454

Find out more about HarperCollins and the environment at **www.harpercollins.co.uk/green**

Health and safety: The publishers, series editor and authors have made every reasonable effort to ensure that the activities in this book are safe when conducted as instructed. However, the publishers, series editor and authors assume no responsibility for any damage or injury caused or sustained while performing the activities in this book to the full extent permitted by the law. Parents, guardians and/or teachers should supervise pupils who undertake the activities in this book.

CONTENTS

INTRODUCTION

WELCOME TO SNAP SCIENCE!

The publication of the new National Curriculum for Science in Primary Schools in England is a wonderful opportunity for subject leaders and teachers to review all aspects of their science provision, including planning, pedagogy, resourcing and assessment. There are several new topics to get to know, a wider range of science enquiry must be planned for, children will need to spend more time learning science outdoors and assessment will no longer be about 'levelling'. All these changes are undoubtedly positive, yet they have big implications for busy, dedicated primary teachers. How can you be sure that the transition to the new Programmes of Study for Science for Key Stages One and Two in your school is smooth and stress free and, most importantly, results in enjoyable, challenging and successful science learning for all children?

Snap Science has been created by a team of leading experts to give subject leaders and teachers the confidence to develop a new scheme of work for science for their school which clearly meets the aims of the new National Curriculum for Science. It covers all the required knowledge, conceptual understanding and the full range of scientific enquiry types identified in the Programmes of Study. Snap Science is a comprehensive and rich resource that will support best practice in science teaching and teacher assessment, whilst encouraging teacher professionalism and autonomy. It has been written by teachers for teachers with full awareness that every class is different but that teachers share similar concerns:

• Am I meeting the requirements of the Programmes of Study?

• Is there clear progression in the science learning in my school?

• Is the level of challenge right for each child in my science lessons?

• Is my subject knowledge secure for all topics?

• Is formative assessment built into every lesson?

• Am I using the best teaching strategies for this topic?

• Are children engaged in productive practical work and meaningful enquiry?

• Do children enjoy their science learning?

• Are the children in my class making good progress in science?

When using Snap Science you can be confident that these concerns are addressed.

Snap Science comprises:

• An online resource kit with a flexible planning tool, editable lesson plans, integrated assets for every lesson (including videos, animations and slideshows), as well as support for assessing and tracking children's progress.

• A printed Teaching Framework per year group with sequenced lesson plans for each topic in the new Programme of Study.

Together these form a tool kit which has been designed to enable and liberate teachers to do what they do best – teach science well! I hope you enjoy using Snap Science.

Jane Turner

THE BEST TEACHING OF PRIMARY SCIENCE

Snap Science has been created to reflect current ideas about best practice in primary science teaching and learning. Although a relatively new subject in primary schools in England, becoming statutory only in 1988, science is a well understood area of the Primary Curriculum. There is a wealth of research evidence and good practice data that has informed the design and content of Snap Science. At the heart of the resource are the following principles:

1 A science scheme of work must embody a clear progression.

2 It is through working scientifically that children develop an understanding of the nature and processes of science and the key scientific knowledge and concepts.

3 Children are curious to find answers to questions about the world around them.

4 Children need to be actively involved in their own learning, to be engaged and reflective.

5 Every child should have the opportunity to achieve in every lesson.

6 Assessment is an integral part of teaching that enables children to understand the purpose of their activities and to improve the quality of their work.

These principles have shaped the overall design of Snap Science, as well as the content and structure of each lesson.

1 Based on the year by year model of the Programmes of Study, the creators of Snap Science have developed a **clear progression framework of the 'big ideas in science'** which has been used to structure the content both within each topic, from year to year and within each year group and module, and to identify any conceptual gaps. This will ensure that children are continually building on their prior learning as they systematically develop their understanding of key ideas and their scientific skills.

2 The creators of Snap Science recognise that **working scientifically**, asking questions and testing ideas against evidence, is the most effective way for children to learn about science. Therefore each lesson **has a clear science enquiry focus**.

3 Every lesson in Snap Science is carefully planned around **a question for children to answer**, either inside the classroom or outside. By ensuring that these questions spark children's curiosity and that they want to find out the answer, lessons are purposeful and result in children gaining a new understanding of the world around them.

4 In each lesson in Snap Science the **learning intention** is designed so that children have a powerful understanding of the skills and understanding they are developing in the lesson. **Success criteria** define the features of the learning intention in the context of the activity so that children can identify what they are aiming for and how well they are doing.

5 Snap Science has been designed to ensure that all children in a class can access and master the lesson's learning intention with each lesson offering **three levels of differentiated task**. These are planned to challenge and extend the learning of all children whilst ensuring that they all achieve the learning intention.

6 Every lesson in Snap Science includes **Assessment for Learning strategies** which enable teachers to find out what children have learned and to use that information formatively.

7 In response to the wealth of evidence that exists about the benefits of children experiencing the natural world first hand, **children learning science outdoors** is a key feature of Snap Science. For each year group there is a module called **Our Changing World** which is designed to be taught in every term, offering children regular opportunities to explore all aspects of their outdoor environment and build up a rich understanding of how it changes over the year.

The most recent Ofsted report[1] into science in primary schools in England outlined three factors which exemplify the best teaching:

- It is driven by determined subject leadership that puts **scientific enquiry at the heart of science teaching** and coupled with **substantial expertise** in how pupils learn science

- It sets out to sustain **pupils' natural curiosity**, so that they are eager to learn the **subject content** as well as develop **the necessary investigative skills**

- It is informed by **accurate and timely assessment** of how well pupils are developing their understanding of science concepts and their skills in analysis and interpretation so that **teaching can respond to and extend pupils' learning**.

The creators of Snap Science have considerable and acknowledged expertise in how children in primary schools learn science well. This experience and knowledge has enabled them to design a resource which embodies the best of current pedagogy and practice, and meets Ofsted's definition of 'the best teaching'.

SCIENCE ENQUIRY

A scheme of work that has **science enquiry at its heart** requires children to learn to use a variety of approaches to answer relevant scientific questions[2]. Each module in Snap Science is made up of a carefully planned series of lessons which will engage children in **the different types of science enquiry identified in the National Curriculum**, where they will use and develop **the necessary investigative skills** and attributes identified for each Key Stage phase. All lessons are stimulated by a question for children to answer, a scientific phenomenon to investigate or a problem to solve. Science Enquiry is the methodology children will use to develop their conceptual knowledge, working in an authentically scientific and purposeful way to collect evidence to find answers to their questions.

SUBJECT KNOWLEDGE CONFIDENCE

Snap Science has been carefully planned to ensure that children use their developing science enquiry skills to build their knowledge of the scientific ideas in a systematic and conceptually appropriate way. Each lesson is designed to explore, value and build on children's prior knowledge so that misconceptions can be addressed and secure understanding developed. The creators of Snap Science are aware that every primary teacher cannot be an expert in all subjects, and have designed the resource to give teachers **subject knowledge confidence**. The sequence of science ideas throughout the resource as a whole, and in each module and lesson, is clear and accurate. The introduction to each module provides teachers with a clear explanation of the science they need to understand and in each lesson key information is highlighted at relevant points.

EXPLORE ACTIVITIES

Stimulating and maintaining **children's natural curiosity** is fundamental to good science teaching and learning. Every lesson in Snap Science starts with an **Explore activity** to excite children's curiosity about a scientific phenomenon and provide a focus for their questions and investigations. The Explore activity is also designed as a **rich formative assessment opportunity** for children to reflect on what they already know, and identify what they need to learn next.

---

[1] Ofsted 2013 Maintaining Curiosity
[2] Goldsworthy, A., Watson, R., Wood-Robinson,V. (2000) *Investigations: Developing Understanding* Hatfield: Association for Science Education
Turner, J., Keogh, B., Naylor, S., Lawrence, L., (2011) *It's Not Fair - Or Is It?* Sandbach: Millgate House Publishers and Association for Science Education

ENQUIRE CHALLENGES

Genuine curiosity leads to **authentic, purposeful science investigation** so each Explore activity is followed by a **differentiated Enquire challenge** where children will collect and analyse data to answer their questions and so develop their scientific understanding and knowledge. Mostly children's science investigations will involve them in **first hand collection, recording and analysis of data**, although sometimes they will use secondary sources of evidence or information to answer questions. In the Enquire challenges children will engage in a **wide range of practical activity** both indoors and outside, using a variety of observation and measuring equipment including data loggers and digital microscopes, everyday items and materials, natural and living things and electrical components. Comprehensive equipment lists are supplied for every module and lesson to help teachers with planning and ensure that **children have as much independence as possible to decide what data to collect to answer their question and how**. Snap Science supports children to work both in groups and alone as appropriate.

SUMMARISE, SHARE, REFLECT

Good science teaching recognises that children need opportunities to **summarise what they have found out, share their findings and reflect on what they have learned**. Each Snap Science lesson has a final **Reflect and Review** activity when children will communicate what they have learned in an appropriate and meaningful way. Being able to summarise understanding is key to developing conceptual knowledge as well as being the vital, final satisfying step in the science enquiry process. **Writing, drawing, speaking, using ICT and mathematical** formats are all important skills in communicating and presenting science and are all developed in Snap Science.

The Reflect and Review part of each lesson also provides an excellent opportunity for children to **self and peer assess their achievements** in the lesson or module, using the success criteria to guide them. How well have I completed the challenge? What do I know now that I didn't know at the beginning of the lesson? What have I learnt to do? What can I do better now than I could at the beginning of the lesson? What do I want to find out next? What do I need to do next to improve the skills I used today?

As children reflect on their own learning, teachers can also assess the progress that they are making. Each lesson in Snap Science includes **guidance for teacher assessment**, indicating where teachers will find evidence of achievement of the learning intention and what that achievement may look like, in the things that children say, do, write or draw. Assessment will therefore be on-going and accurate – focusing firmly on progress in conceptual knowledge plus data collection and analysis skills. Assessment will also be formative, supporting teachers and children to identify the next steps in learning, and to keep moving forward.

THE NEW PROGRAMMES OF STUDY FOR SCIENCE

In September 2013 the Department for Education published a new Primary National Curriculum including a new Programme of Study for Science. Implementation begins September 2014 with the first sample tests reporting on national standards against the new POS in 2016.

What is different about the new National Curriculum for Science? What challenges and opportunities do these changes present to primary teachers? How will Snap Science help teachers to meet them?

As with any curriculum it is important to know and understand its aims. It is the aims that explain what the education described should do. In the case of the new National Curriculum for Science the aims are reflective of widely and long held views about the values and purposes of primary science education.

The National Curriculum for Science aims to ensure that all children:

1 Develop **scientific knowledge and conceptual understanding** – through the specific disciplines of biology, physics and chemistry,

2 Develop understanding of the **nature, processes and methods of science** through different types of science enquiry that help children to answer scientific questions about the world around them, and

3 Are equipped with the scientific knowledge required to understand the **uses and implications** of science, today and for the future.

1 The Programmes of Study contain a **sequence of knowledge and concepts** on a year by year basis. Although it is not compulsory for schools to follow this sequence, the blocks of knowledge and concepts have been arranged to ensure progress in the big ideas of science to ensure that children develop secure understanding. Most of the content will be familiar to teachers, but some topics have been broadened and extended, and some are introduced at different times.

- The plant and animal biology content is significantly increased for every year group, with much more focus on children getting to know and to classify living things in their local and wider outdoor environment.

- A wider range of plant and animal life cycles is included.

- Evolution and inheritance is now included at Y6.

- Simple digestion in humans is included at Y3.

- Seasonal changes is now included at Y1 including day length.

- Chemistry at Y3 now includes fossils.

- States of matter is now in Y4, mixing and changing materials is all in Y5.

- There is no chemistry content specified for Y6.

- These is no requirement to cover electricity, light and sound, and forces and movement in KS1.

- Mechanisms are included at Y6.

To enable teachers to confidently plan to teach the new National Curriculum, Snap Science is organised into a series of modules per year group, based on the topics in the new Programmes of Study, and the year groups in which they occur. All the required science content is fully covered to make sure that learning is deep. Each module is made up of a sequence of lessons which has been carefully planned by subject experts to ensure that new ideas are only introduced once understanding of lower-order content is secure. Although in most lessons children will engage in first hand practical activity, each lesson is also richly supported by additional assets including film clips, images and interactive resources to exemplify and illustrate concepts and ideas.

Because the creators of Snap Science know that primary schools are organised in many different ways the interactive planning tool will facilitate flexible whole school planning, whilst still ensuring coherent progression of knowledge and understanding.

2 In the new National Curriculum **the nature, processes and methods of science** are organised into three sections for KS1, lower and upper KS2 under the title **working scientifically**. These sections of the Programmes of Study identify the progress children should make in all areas of science enquiry from asking questions, planning and carrying out investigations, presenting data, making and communicating conclusions and evaluating results.

WORKING SCIENTIFICALLY

However a difference in the new Programmes of Study is that working scientifically, although described separately, is not to be taught as a separate strand. The creators of Snap Science fully support this embedded approach as it reflects accurately the scientific approach and ensures that science lessons are purposeful and lead to the learning of science concepts. Each lesson in Snap Science is planned to meet a biology, chemistry or physics learning intention by Working Scientifically. Teachers will therefore be able to track progression in all aspects of the Programmes of Study.

The new Programmes of Study require that children should learn to use a variety of approaches to answer scientific questions, as well as fair testing which has become an over and often incorrectly used method in primary classrooms[3]. The creators of Snap Science know that different questions lead to different types of enquiry. Teachers can be confident that the starting points in Snap Science will require children to use the recommended different types of science enquiry to answer questions and so develop their understanding about which is the best method to use to answer a question. In each lesson in Snap Science the enquiry strategy that children will use is clearly identified. Enquiry strategies include those recommended in the new National Curriculum:

- **Observing over time** – when children observe or measure how one variable changes over time
- **Identifying and classifying** – when children identify and name materials and living things and make observations or carry out tests to organise them into groups
- **Looking for patterns** – when children make observations or carry out surveys of variables that cannot be easily controlled and look for relationships between two sets of data
- **Comparative and fair testing** – when children observe or measure the effect of changing one variable when controlling others as far as possible
- **Answering questions using secondary sources of evidence** – when children answer questions using data or information that they have not collected first hand

As well children will:

- **Use models** – to develop or evaluate a model or analogy that represents a scientific idea, phenomenon or process

---

[3] Turner, J., Keogh, B., Naylor, S., Lawrence, L., (2011) *It's Not Fair – Or Is It?* Sandbach: Millgate House Publishers and Association for Science Education

3 Teachers sometimes find daunting the idea of helping children to understand the **uses and implications of science**, citing their own lack of confidence about complex scientific developments and issues. The creators of Snap Science recognise that this aim is challenging and have developed a straightforward two-fold strategy to support teachers to develop children's scientific literacy.

- Firstly the **context of the lesson must make sense and matter to children**. They need to see the relevance of a scientific question or concept to their own lives. In every lesson children should be able to explain the importance of the question they are answering, and how the science connects to their own lives. In Snap Science the approach to this is pragmatic and manageable, using question and stimulus Explore activities to focus children's powerful curiosity about the world around them. At the end of every lesson in Snap Science children should be able to explain or demonstrate how they have answered the question and what they have learnt. Sometime children will do this via a **technology** activity – applying their scientific knowledge and understanding to make an artefact or system that solves a problem. Sometimes they will use argument, debate or persuasive writing to show how a science explanation has helped them to understand or make informed personal decisions about something that involves science, such as health, diet, use of energy resources, or human impact on the natural environment.

- Secondly children must learn that all **understanding in science depends on the evidence that has been used to answer a question** and that working scientifically involves evaluating the quality of that evidence and the conclusions that have been drawn from it. In every lesson children should be able to explain what evidence they have used to answer a question and to evaluate honestly the reliability and validity of that evidence. In Snap Science children are supported to use discussion and argument to evaluate their own data, methods and conclusions, as well as those of others, in the classroom and beyond. They will also be helped to recognise how improvements in evidence collection and analysis techniques have led to ideas about science changing over time.

ASSESSING THE NEW NATIONAL CURRICULUM FOR PRIMARY SCIENCE

Alongside new Programmes of Study for primary science, a new assessment model will be introduced, where attainment will no longer be tracked and reported against level descriptors, but instead children's 'mastery' of the matter, skills and processes of the Programme of Study will be assessed. The expectation is that most children will achieve 'mastery' of the full programme of study. Sample tests in Science to provide national monitoring of standards will take place every two years, but involving a small number of pupils, with no individual pupil or school Science attainment data being reported. Schools are now free to decide how to track the progress that children make through KS1 and 2 against the Programmes of Study.

What are the implications of removing levels in primary science? What challenges and opportunities does this change present to primary teachers? How will Snap Science help teachers to meet them?

Levels, particularly when used to design SAT questions, had the effect of constraining science learning in two ways: firstly the curriculum became narrowed to what could be tested in a replicable, pen and paper method and therefore understanding of what achievement looked like came to be understood in terms of these narrow tasks; and secondly they became organising models for planning, with teachers differentiating lessons according to artificial notions of children 'levelness'. [4]

---

[4] Harlen (2012), Developing Policy, Principles and Practice in Primary Science Assessment: Report from a Working Group; London: Nuffield Foundation

The removal of level descriptors means that the **relationship between the science that children are taught and the science that is assessed will be much stronger**. Confident teacher assessment is vital and the creators of Snap Science have ensured that **effective formative assessment strategies** are used in every lesson.

- Each lesson has a clear science **Learning Intention** which all children are expected to achieve or exceed, with **Success Criteria** to exemplify what success will look.

- Differentiation is by access, with each lesson beginning with an **Explore** activity to enable children and their teacher to assess prior understanding and identify which level of challenge to take. Teachers can annotate planning to reflect this.

- The **Enquire** part of the lesson includes a choice of three challenges which will ensure that all children can work appropriately towards achieving the learning intention. Differentiation in the challenges is based on a model of progression in science learning which supports children to become more independent and autonomous, systematic, precise and evaluative, and to increasingly use their scientific knowledge in their explanations. This means that grouping in Snap Science lessons is flexible, dependent on the level of skill, knowledge and understanding that each child demonstrates in the Explore activity, and the level of support and challenge that is appropriate for them in each lesson. Children should be encouraged, with teacher support, to choose for themselves the right challenge to complete to achieve the learning intention.

- The final stage of each lesson is a **Reflect and Review** activity where children summarise what they have learnt and use the success criteria to assess their success and identify next steps. Assessment evidence from each lesson should be used formatively to determine appropriate next steps for individuals and groups of children.

Without levels to track progress against teachers will need to find other ways of monitoring children's progress and reporting this to parents, secondary schools and external bodies such as Ofsted. The creators of Snap Science recognise that this represents a challenge to schools and have referred to recognised sources of good practice[5] to design a manageable process for tracking progress in science.

- **Formative assessment evidence from each lesson,** including children's work, the feedback that is given and responded to and any additional observation notes that the teacher makes, is used to track progression and to enable teachers to make confident summative judgments of attainment when required. Comparison of evidence outcomes from different children and between classes and school is used to **moderate teacher assessment judgements**.

- Supporting digital assessment tasks on Collins Connect such as quizzes or short activities can be used to **check children's understanding** of specific concepts or facility with particular working scientifically skills.

- Either at an identified point during a module or at the end of a module a teacher reviews any observation notes on a child, their written work, their self-assessment judgements, and their answers to any additional activities they have completed to ascertain if they **have not yet achieved, have achieved, or have achieved and exceeded the expected outcomes** for that part of the Programme of Study. Teachers **record the judgements in an on-going digital progress tracker on Collins Connect** which contains a summary of the knowledge outcomes for that module including an on-going Working Scientifically tracker.

- There are no 'end of module tests' nor artificial interim (sub) levels of achievement. It is assumed that high quality formative assessment which influences planning for individual children will lead

---

[5] Harlen (2012), Developing Policy, Principles and Practice in Primary Science Assessment: Report from a Working Group; London: Nuffield Foundation

to an excellent match of task and challenge and that teachers and children will recognise when the lesson intention has been achieved.

- At the end of each Key Stage teachers use the digital progress trackers on Collins Connect to judge **whether or not each child has achieved the designated learning outcomes for the Key Stage** in the main components of the National Curriculum.

- At the end of the Key Stage **individual pupil records are aggregated from each class,** for the school as a whole and for particular groups.

SNAP SCIENCE COMPONENTS

THE TEACHING FRAMEWORK

- The printed Teaching Framework is organised into a series of modules, based on the topics in the new Programmes of Study. The new Programme of Study for Science is divided into 4 topics per year in KS1 and 5 per year in KS2. Some topics have considerably more content than others. Traditional half termly topic planning for science will clearly no longer be appropriate. The modules in Snap Science have been organised to reflect the content of the POS for that year and are not all of identical length. Teachers should plan to teach all the complete modules over a year, regularly fitting in lessons from the Our Changing World module.

- Each module begins with an introduction that provides background information on the topic at an appropriate level for non-specialist teachers as well as advice on the misconceptions or alternative conceptions that research indicates that children frequently develop as they make sense of the world around them.

- Each module then contains a sequence of lesson plans. Snap Science is designed around the principle that science should be taught at least once a week throughout the year, as Ofsted recommends.

- The modules are divided into 'core' and 'enrichment' lessons. The 'core' lessons cover all the objectives from the Programme of Study. The 'enrichment' lessons provide extra breadth and depth for the topic.

A SNAP SCIENCE LESSON

Every lesson begins with a **question** – providing a focus for children to explore and think about.

The 'C' or 'E' symbol indicates whether it is a 'core' or 'enrichment' lesson

Key vocabulary highlights important technical and descriptive language that should be used as part of the lesson.

Resources lists all the materials you will need.

The **learning intention** establishes a clear aim for what children should learn over the course of the lesson. Teachers will not always want to share this with children at the beginning of the lesson.

The **Scientific enquiry type** is highlighted where appropriate

The **Explore** section begins the lesson. It introduces the science phenomena, sparking curiosity and enabling children to share their previous understanding and generate questions to investigate.

The **Enquire** section is where children will answer these questions. It is divided into three levels of differentiated challenge to ensure all children can access and master the lesson's learning intention.

The **lesson summary** gives a quick overview of the main activity in the lesson, so you can see at a glance what you will be covering.

Each lesson links directly to the **Programmes of Study** and the **Working Scientifically** criteria.

Success criteria for each lesson are written in child speak. They exemplify what successful attainment of the learning intention looks like.

Prompt questions are included throughout as ways of eliciting or developing children's understanding

---

MODULE 2

ROCK DETECTIVES

C **LESSON 3: HOW ARE ROCKS USED AROUND OUR SCHOOL?**

Key vocabulary:
rock, sandstone, granite, chalk, limestone, marble, concrete, slate, brick, clay, stone, tile, roof, floor, pavement, wall

Resources:
No extra resources

LESSON SUMMARY:
In this lesson children will identify where and how rocks are used in their local environment. By the end of the lesson children will be able to identify a variety of contexts where rocks have been used and explain why their properties make them particularly suited to the job that they are doing.

National curriculum links:
Compare and group together different kinds of rocks on the basis of their appearance and simple physical properties

Learning intention:
To recognise where and how rocks are used and explain how their properties make them suitable for their purpose

Scientific enquiry type:
Grouping and classifying

Working scientifically links:
Gathering, recording, classifying and presenting data in a variety of ways to help in answering questions

Success criteria:
• I can identify different types of rocks found around the school.
• I can describe how they are being used.
• I can explain why their properties make them useful for this purpose.

EXPLORE:
Explore key questions with children, reviewing their prior learning from the previous two lessons.
Ask: *What rocks do you know of already? What are their properties? Which do you think is the hardest? Which the softest? How might they be used?*
Play 'Rock or not' (Interactive 1), an interactive whiteboard game featuring images of different buildings, structures, street furniture and so on. Encourage the children to discuss the rocks they can see pictured. They then drag and drop images into 'Rock' or 'Not' bins.

ENQUIRE:
Tell the children that they are going to be Rock Detectives, and that their challenge is to find out how rocks are used around school and in places further afield. The challenges are differentiated by the level of detail required in their survey, the support given in recording the investigation and the manner of presentation. The challenges are presented on the Challenge slides to be displayed on the board, or printed out and placed in the centre of the table.

Challenge 1: Children identify and sort rocks in use in the environment around them
Display Rocks in our environment (Slideshow 1) or print out and laminate the photos from the slideshow to give to children. They review, identify, sort and group the images of buildings, structures and objects, some made of rock or rock derivatives, others made of other materials, some of which look like rock, such as plastic garden centre pots.
Ask: *Why was rock used to make some of the objects in the photographs? What types of rocks would be best to use?*

Challenge 2: Children carry out a survey of the rocks found in their local school environment.
Organise children into groups and give them their Rock Checklist (Resource sheet 1). You may wish to group the children and send them to different locations in order to contribute to a more extensive survey of the school and grounds; for example, group 1 corridor, group 2 classrooms, group 3 entrance and school office, group 4 school field, group 5 school gate and roadway. The children name the objects made of rock, suggest what rock they think has been used (using the checklist) and explain why.
Provide them with a table (Resource sheet 2) in which to record their survey findings (either on paper or using iPads) and remind them to act as thorough Rock Detectives, looking carefully and in detail as they complete their task. When they return to the classroom, ask them to review what they have found out.

Prompts are provided throughout the lesson plan to highlight when there is a **related asset** for you to use.

Key information is provided throughout, providing helpful tips and scientific background knowledge.

Opportunities to link lessons to other subject areas are flagged as appropriate.

White space for teachers to make notes and changes to lesson plans. Snap Science is designed to be supportive not prescriptive!

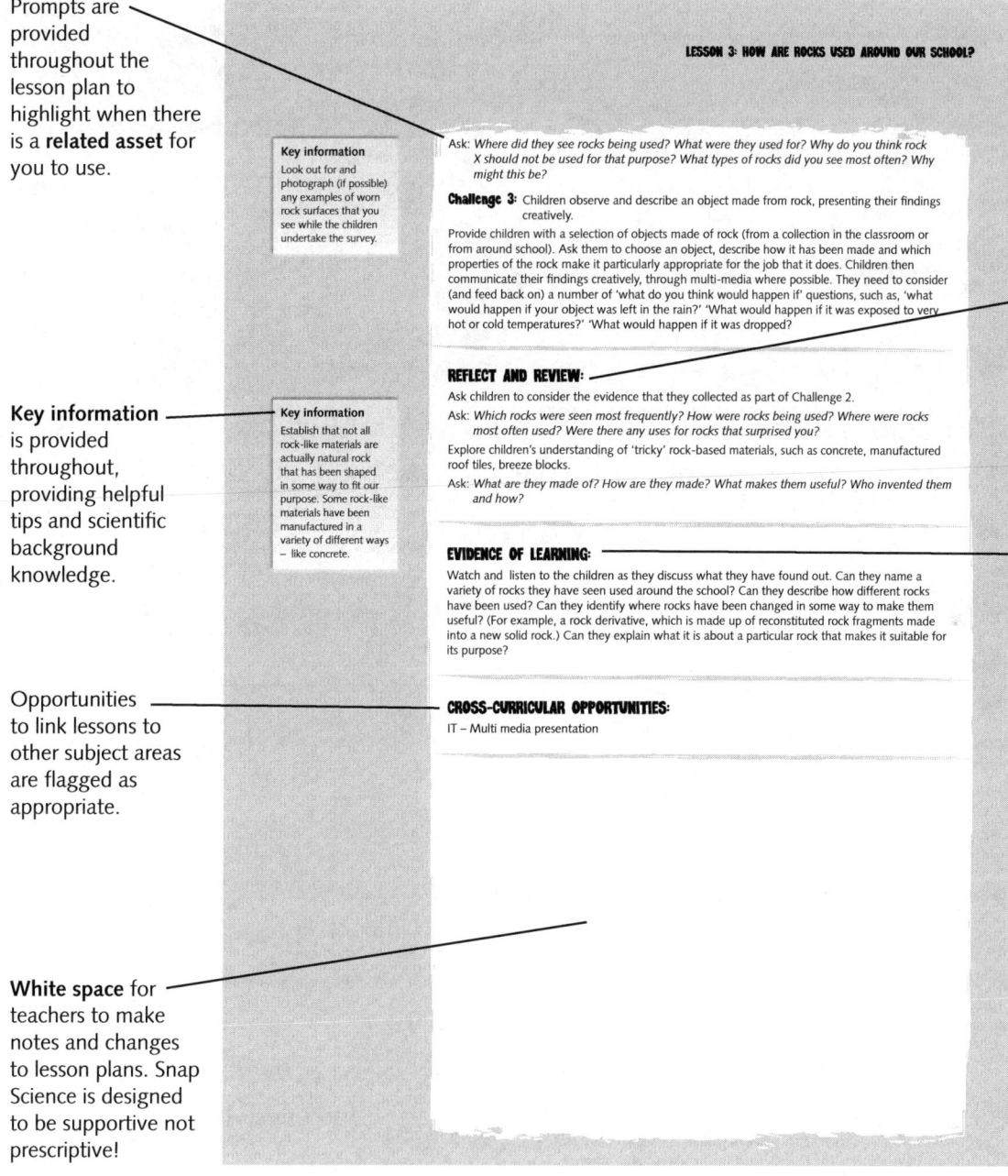

LESSON 3: HOW ARE ROCKS USED AROUND OUR SCHOOL?

Key information
Look out for and photograph (if possible) any examples of worn rock surfaces that you see while the children undertake the survey.

Ask: *Where did they see rocks being used? What were they used for? Why do you think rock X should not be used for that purpose? What types of rocks did you see most often? Why might this be?*

Challenge 3: Children observe and describe an object made from rock, presenting their findings creatively.

Provide children with a selection of objects made of rock (from a collection in the classroom or from around school). Ask them to choose an object, describe how it has been made and which properties of the rock make it particularly appropriate for the job that it does. Children then communicate their findings creatively, through multi-media where possible. They need to consider (and feed back on) a number of 'what do you think would happen if' questions, such as, 'what would happen if your object was left in the rain?' 'What would happen if it was exposed to very hot or cold temperatures?' 'What would happen if it was dropped?

Key information
Establish that not all rock-like materials are actually natural rock that has been shaped in some way to fit our purpose. Some rock-like materials have been manufactured in a variety of different ways – like concrete.

REFLECT AND REVIEW:

Ask children to consider the evidence that they collected as part of Challenge 2.

Ask: *Which rocks were seen most frequently? How were rocks being used? Where were rocks most often used? Were there any uses for rocks that surprised you?*

Explore children's understanding of 'tricky' rock-based materials, such as concrete, manufactured roof tiles, breeze blocks.

Ask: *What are they made of? How are they made? What makes them useful? Who invented them and how?*

EVIDENCE OF LEARNING:

Watch and listen to the children as they discuss what they have found out. Can they name a variety of rocks they have seen used around the school? Can they describe how different rocks have been used? Can they identify where rocks have been changed in some way to make them useful? (For example, a rock derivative, which is made up of reconstituted rock fragments made into a new solid rock.) Can they explain what it is about a particular rock that makes it suitable for its purpose?

CROSS-CURRICULAR OPPORTUNITIES:

IT – Multi media presentation

The **Reflect and Review** section provides the opportunity for children to consolidate the key learning from the lesson.

Evidence of Learning provides assessment guidance for teachers, indicating the kinds of things children might say, write, draw or do to demonstrate they have achieved or exceeded the learning intention.

THE ONLINE TOOLKIT

- You can access all of the lesson sequences, lesson plans and related assets online via the **Collins Connect platform**. There is a discrete area for each year group, with support for planning, teaching and tracking progression.

- Simply select your year group using the tabs along the top. Using the drop down lists, choose the module you would like to teach and then a lesson from the sequence. Click on the lesson to reveal the lesson plan and related assets.

- **The lesson plan is provided in an editable format**, so you can adapt and customise the content for your class.
- A **range of digital assets** are provided to help make your science lessons rich, lively and engaging.
- All of the assets for the lesson are clearly labelled, with the **asset type highlighted** so you can easily navigate to each one.

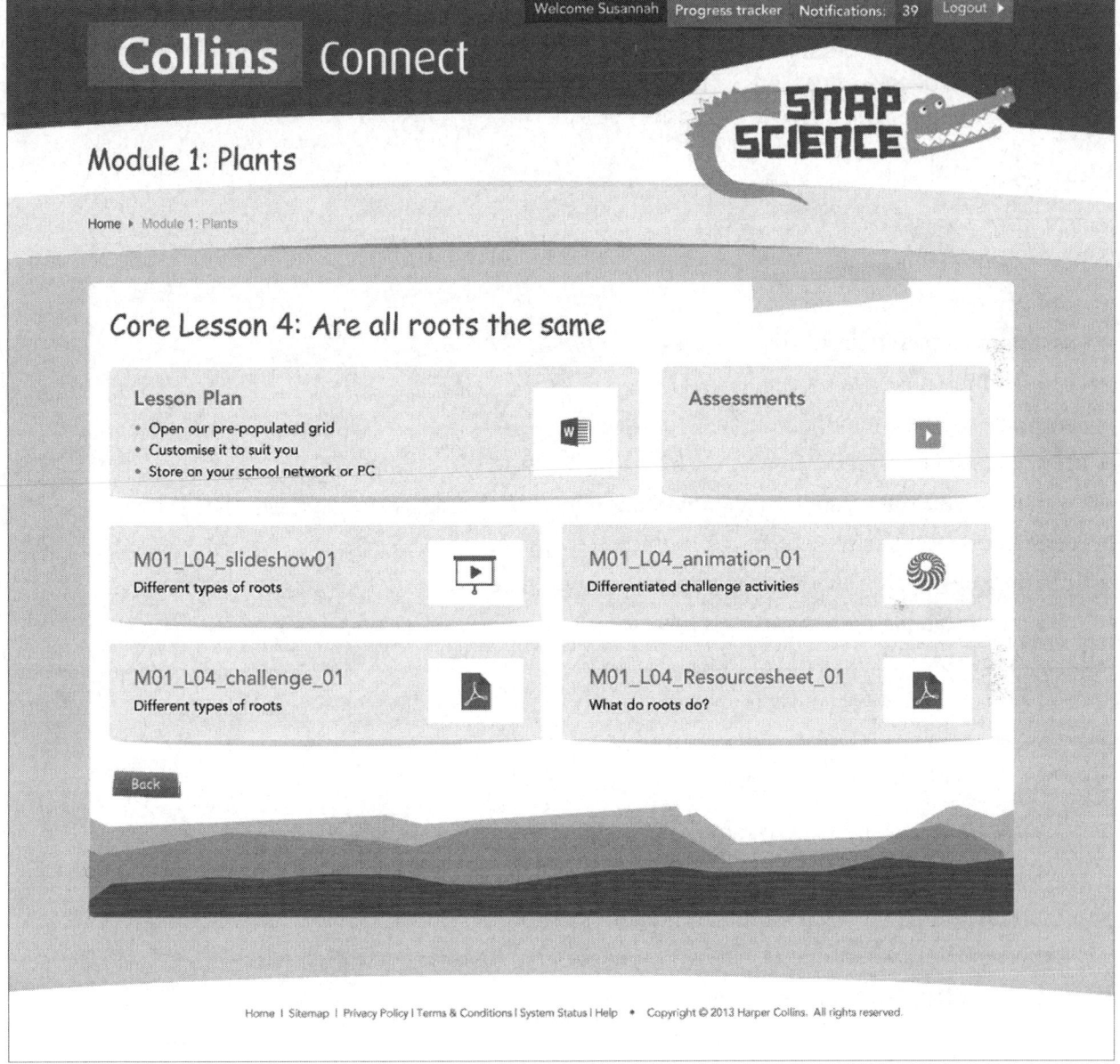

PROGRESSIVE SUCCESS CRITERIA FOR DATA COLLECTION AND ANALYSIS SKILLS

In every lesson in Snap Science success criteria are used to break down the learning intention to help children to identify what the steps to success are. Often success criteria refer to the specific skills of collecting, presenting and interpreting data, for example *I can record my observations in a table*.

For more detailed understanding of what success looks like in these data collection and analysis skills, teachers and children can use the following sets of progressive success criteria. They should be used formatively to ensure that children are motivated to reach goals that are specific, within reach and offer some degree of challenge; also summatively by children and teachers to assess the level of success and therefore plan the next steps.

Measuring
- I can take non-standard measures
- I can choose equipment to take measurements
- I can use standard measures
- I can start measuring from zero
- I can read the numbers on a scale
- I can estimate between the numbers on a scale
- I can convert between different units of metric measure e.g. m/km
- I can decide how accurate to make my measurements
- I can choose equipment with an appropriate data range
- I can recognise that a series of measurements should be made
- I can select suitable ranges and intervals
- I can make sure I collect enough data
- I can repeat readings to check my results
- I can use compound units for speed

Collecting data
- I can record my results in a tally chart
- I can construct a tally chart
  - I can draw a simple chart with three columns
  - I can put the different things I am counting in separate rows
  - I can count in 5's making tally marks
  - I can add up the frequency in the final column
- I can record my results in a table
- I can construct a table
  - I can draw a simple table using two lines
  - I can write what I am changing in the first column
  - I can write what I am measuring in the second column
  - I can include the units
- I can add extra columns to my table for repeat readings
- I can calculate the average
- I can add a column for the average reading

Presenting results

- I can present my results as a pictogram
- I can construct a pictogram
  - I can decide what picture to use
  - I can include a key
- I can present my results as a bar graph
- I can construct a bar graph
  - I can draw the axes using a ruler
  - I can label the axes
  - I can plot what I have changed (the independent variable) on the horizontal x axis and what I have measured in order to collect the results (the dependent variable) on the vertical y axis
  - I can choose an appropriate scale for the y axis
  - I can choose a scale that goes up to a suitable number
  - I can choose a scale that increases by regular intervals
  - I can draw bars of the correct height
- I can construct a bar line graph
- I can present my results as a line graph
- I can construct a line graph
  - I can draw the axes using a ruler
  - I can label the axes
  - I can plot what I have changed (the independent variable (what is changed) goes on the horizontal x axis and what I have measured in order to collect the results (the dependent variable) on the vertical y axis
  - I can choose an appropriate scale for the axes
  - I can choose a scale that goes up to a suitable number
  - I can choose a scale that increases by regular intervals
  - I can plot points accurately
  - I can join the points
- I can choose which type of graph to use

Interpreting data

- I can answer questions about one value in a pictogram by counting the pictures
- I can answer questions about one value in a block graph or line graph by using the numbers on the y axis
- I can answer comparative questions about two or more values in a pictogram or block graph
- I can ask questions using the points on a line graph
- I can answer questions using the line on a line graph
- I can read between the points on a line graph
- I can use the line graph to make predictions outside the data collected
- I can use a pictogram, bar graph or bar line graph to make statements about quantity/ measurements or frequency
- I can use a line graph to make statements about patterns between two sets of continuous data

SUCCESS CRITERIA FOR DIFFERENT APPROACHES TO SCIENCE ENQUIRY

These success criteria can be used by teachers and children to plan different aspects of scientific enquiry

Comparative test – comparing one thing with another to investigate a difference between them

- I can recognise when to use a comparative test to answer a question
- I can identify properties or features to compare
- I can choose how to observe or measure the feature I am comparing
- I can decide what equipment to use
- I can decide how to collect and present data
- I can write a clear question for my comparative test
- I can explain how I made my results reliable

Fair test – controlled testing to investigate the impact of changing one variable on another

- I can recognise when to use a fair test to answer a question
- I can identify (independent) variables to change
- I can identify (dependent) variables to observe or measure
- I can choose the (independent) variable to change
- I can choose the (dependent) variable to measure
- I can identify the variable to keep the same (control)
- I can identify which variables cannot be controlled
- I can decide what equipment to use
- I can decide how to collect and present the data
- I can write a clear question for my fair test
- I can explain how I will make the test fair

Noticing patterns – when children make observations or carry out surveys of variables that cannot be easily controlled to investigate relationships between two sets of data

- I can recognise when to look for patterns to answer a question
- I can identify patterns I could investigate
- I can choose two variables to observe or measure
- I can decide what equipment to use
- I can decide on sample size
- I can decide how to collect and present my data
- I can write a clear question for my pattern seeing investigation

Observing changes over time – when children observe or measure how one variable changes over time to investigate how change takes place

- I can recognise when to make observations over time to answer a question
- I can identify variables to observe or measure
- I can decide how often to make observations or take measurements
- I can decide how long to continue to make observations or take measurements
- I can decide what equipment to use
- I can decide how to collect and present my data
- I can write a clear question for my observing over time investigation

Grouping and classifying things – when children identify materials and living things and make observations or carry out tests to identify similarities and differences and organise them into groups

- I can recognise when to identify and classify objects, materials or living things to answer a question
- I can identify observable differences and similarities to observe or measure
- I can decide what equipment to use
- I can decide how to collect and present data
- I can identify behavioural differences and similarities to observe or measure
- I can identify simple tests to carry out to identify differences and similarities
- I can write a clear question for my identifying and classifying investigation

Finding things out using secondary sources of information – when children answer questions using data or information that they have not collected first hand

- I can recognise when to use secondary data to answer a question
- I can suggest how to find things out
- I can decide what information sources to use
- I can decide how to collect and present data
- I can write a clear question for my research using secondary data investigation

PROGRESSION CHARTS

The National Curriculum Programme of Study for Science describes a sequence of knowledge and concepts, processes and methods. This sequence of knowledge and concepts is arranged as progressive blocks of key ideas in biology, chemistry and physics, alongside a progression in the skills of working scientifically.

The conceptual ideas in Biology, Chemistry and Physics build on each other and children need to develop a strong understanding of each set of ideas in order for the next set to make sense and for them to make progress. The Programme of Study is set out year by year for Key stages 1 and 2 but each science topic is not covered in every year. It is therefore important that teachers and children know where each block of ideas fits into the overall sequence.

In the Snap Science Progression Charts the key ideas within Biology, Chemistry and Physics in the National Curriculum are arranged to show how they are related to each other and how one idea builds on another. The National Curriculum statements have been edited into key ideas statements. The source of each key idea is identified by the year group and the Programme of Study topic heading. Some additional statements have been added to make important links between ideas.

Working Scientifically is taught throughout KS1 and 2, embedded within the content of Biology, Chemistry and Physics. The National Curriculum Programme of Study for Working Scientifically outlines the practical scientific methods, processes and skills that children must be taught to use, divided into three two-year blocks. In every lesson in Snap Science children will use their developing science enquiry skills to answer scientific questions. The Snap Science Progression Chart for Working Scientifically exemplifies the progression in these skills in the key areas of raising questions and planning, collecting and presenting data, drawing and evaluating conclusions.

This progression underpins the sequence of teaching and learning in each Snap Science module and between year groups.

BIOLOGY: progression of ideas through KS1 and 2

LIFE PROCESSES

STRUCTURE AND FUNCTION

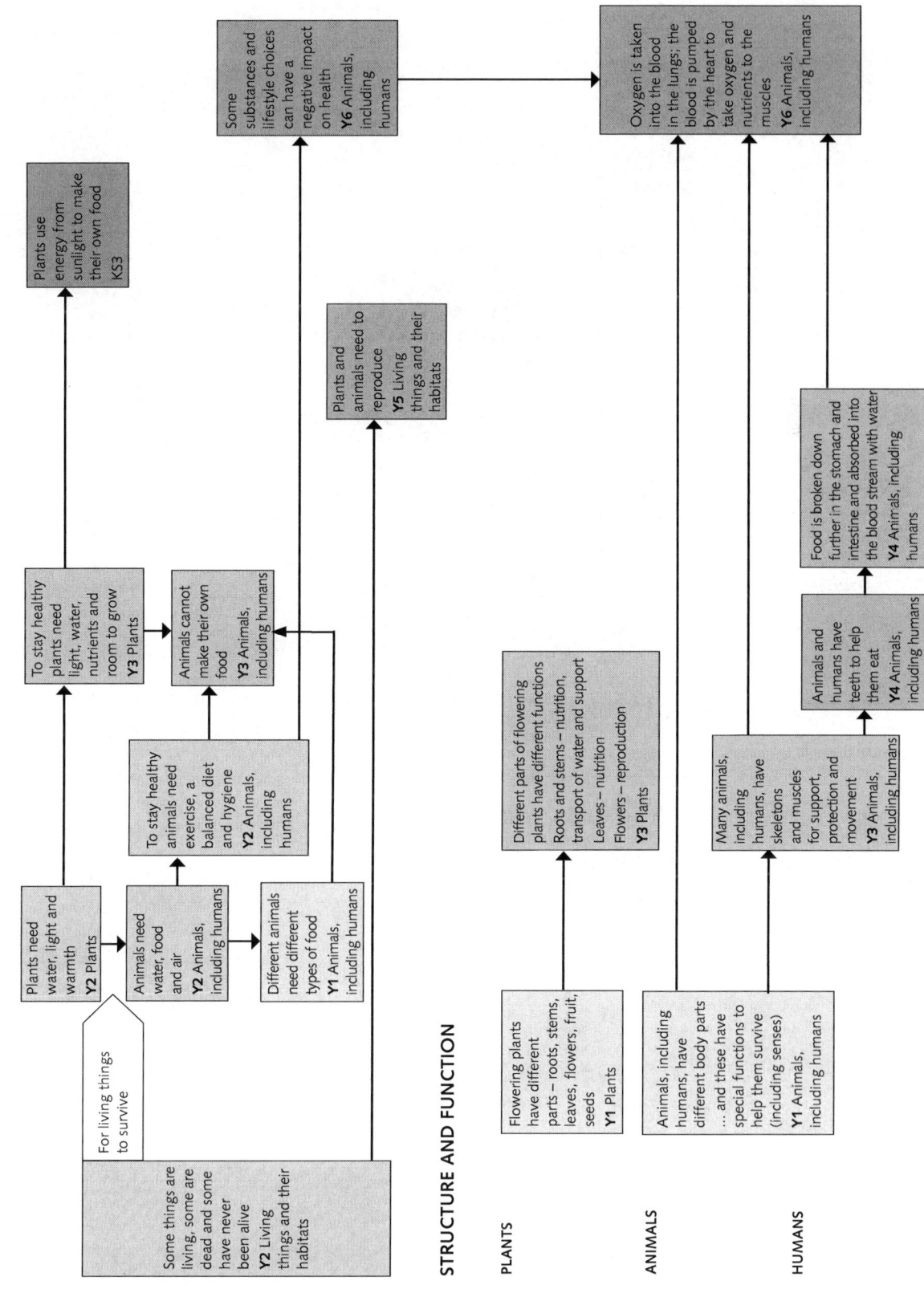

CLASSIFICATION

Identifying and classifying increasing range from the familiar to the unfamiliar

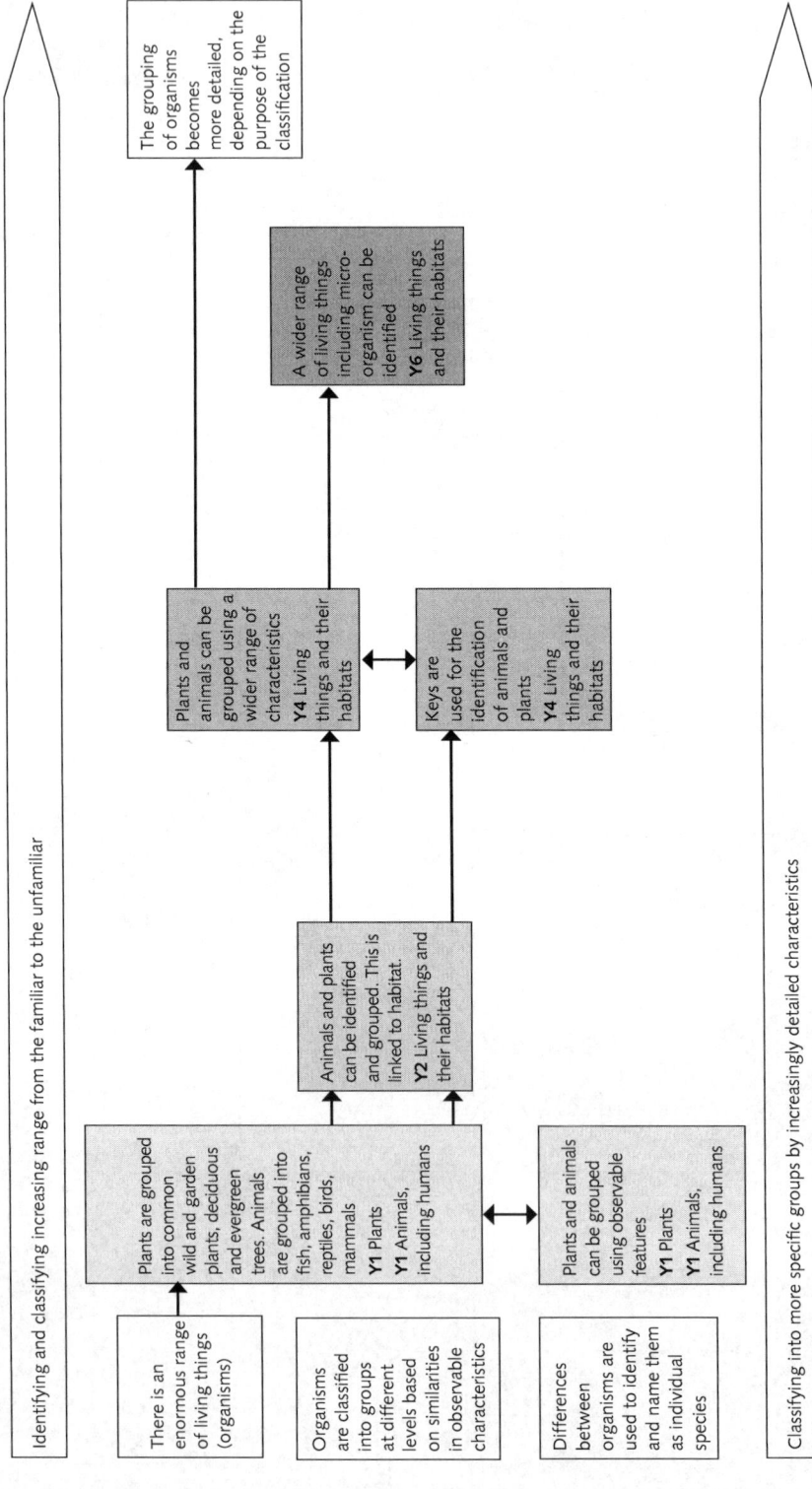

The grouping of organisms becomes more detailed, depending on the purpose of the classification

A wider range of living things including micro-organism can be identified **Y6** Living things and their habitats

Plants and animals can be grouped using a wider range of characteristics **Y4** Living things and their habitats

Keys are used for the identification of animals and plants **Y4** Living things and their habitats

Animals and plants can be identified and grouped. This is linked to habitat. **Y2** Living things and their habitats

There is an enormous range of living things (organisms)

Plants are grouped into common wild and garden plants, deciduous and evergreen trees. Animals are grouped into fish, amphibians, reptiles, birds, mammals **Y1** Plants **Y1** Animals, including humans

Organisms are classified into groups at different levels based on similarities in observable characteristics

Plants and animals can be grouped using observable features **Y1** Plants **Y1** Animals, including humans

Differences between organisms are used to identify and name them as individual species

Classifying into more specific groups by increasingly detailed characteristics

LIFE CYCLES

INTERDEPENDENCE

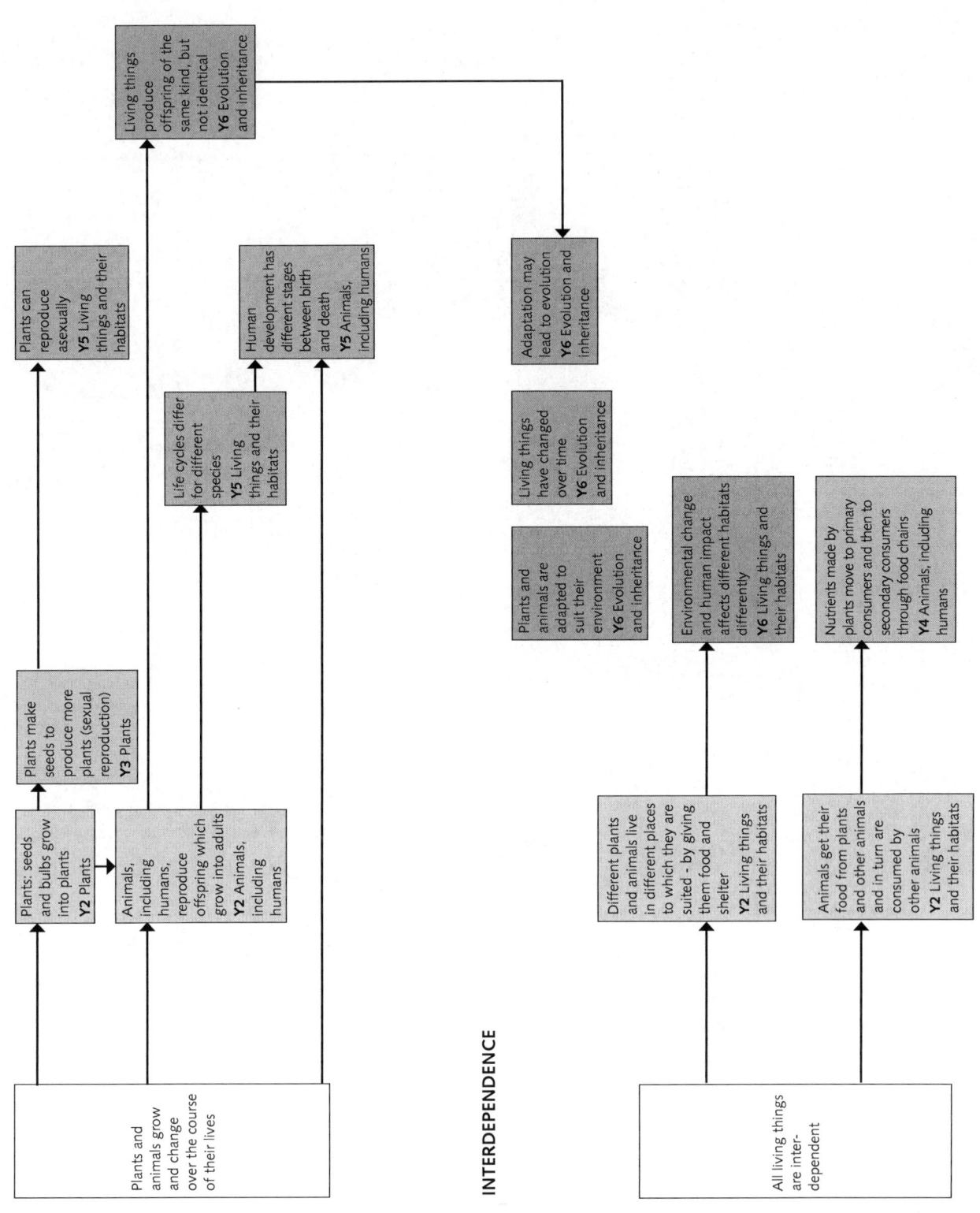

CHEMISTRY: progression of ideas through KS1 and 2

MATERIALS

DESCRIBING AND USING MATERIALS

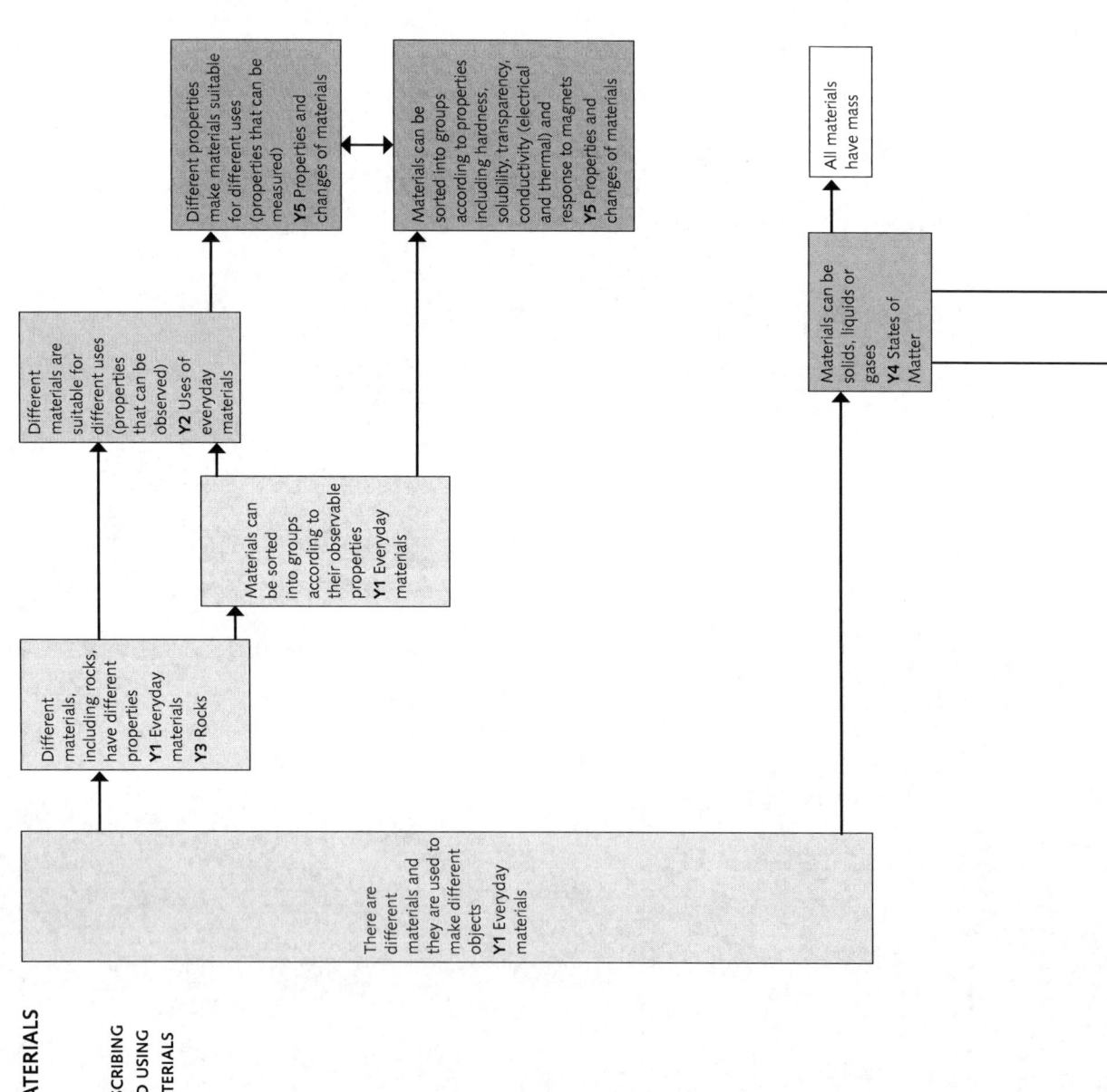

There are different materials and they are used to make different objects
Y1 Everyday materials

Different materials, including rocks, have different properties
Y1 Everyday materials
Y3 Rocks

Materials can be sorted into groups according to their observable properties
Y1 Everyday materials

Different materials are suitable for different uses (properties that can be observed)
Y2 Uses of everyday materials

Different properties make materials suitable for different uses (properties that can be measured)
Y5 Properties and changes of materials

Materials can be sorted into groups according to properties including hardness, solubility, transparency, conductivity (electrical and thermal) and response to magnets
Y5 Properties and changes of materials

Materials can be solids, liquids or gases
Y4 States of Matter

All materials have mass

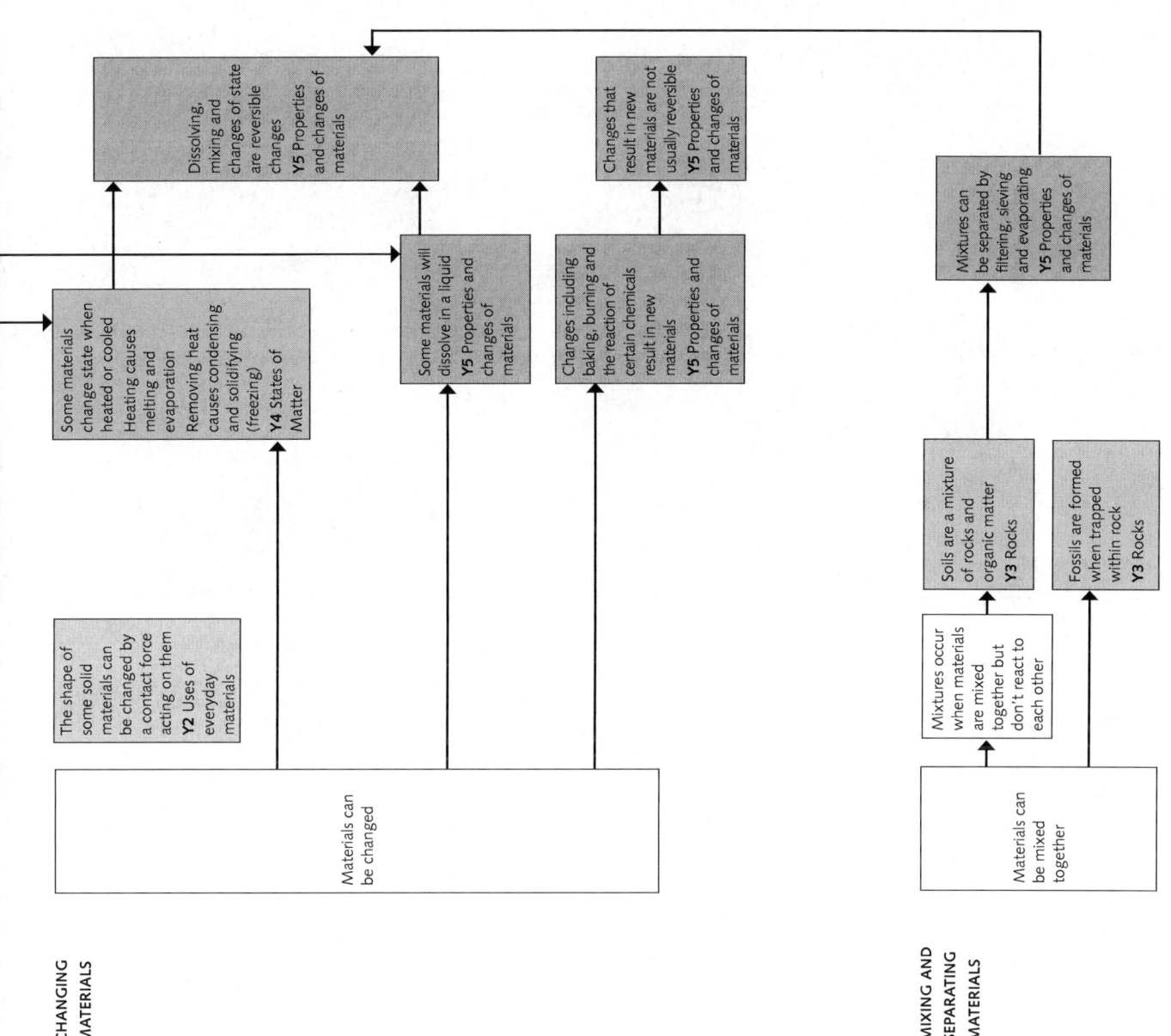

CHANGING MATERIALS

Materials can be changed

The shape of some solid materials can be changed by a contact force acting on them **Y2** Uses of everyday materials

Some materials change state when heated or cooled
Heating causes melting and evaporation
Removing heat causes condensing and solidifying (freezing) **Y4** States of Matter

Dissolving, mixing and changes of state are reversible changes **Y5** Properties and changes of materials

Some materials will dissolve in a liquid **Y5** Properties and changes of materials

Changes including baking, burning and the reaction of certain chemicals result in new materials **Y5** Properties and changes of materials

Changes that result in new materials are not usually reversible **Y5** Properties and changes of materials

MIXING AND SEPARATING MATERIALS

Materials can be mixed together

Mixtures occur when materials are mixed together but don't react to each other

Soils are a mixture of rocks and organic matter **Y3** Rocks

Fossils are formed when trapped within rock **Y3** Rocks

Mixtures can be separated by filtering, sieving and evaporating **Y5** Properties and changes of materials

PHYSICS: progression of ideas through KS1 and 2

LIGHT

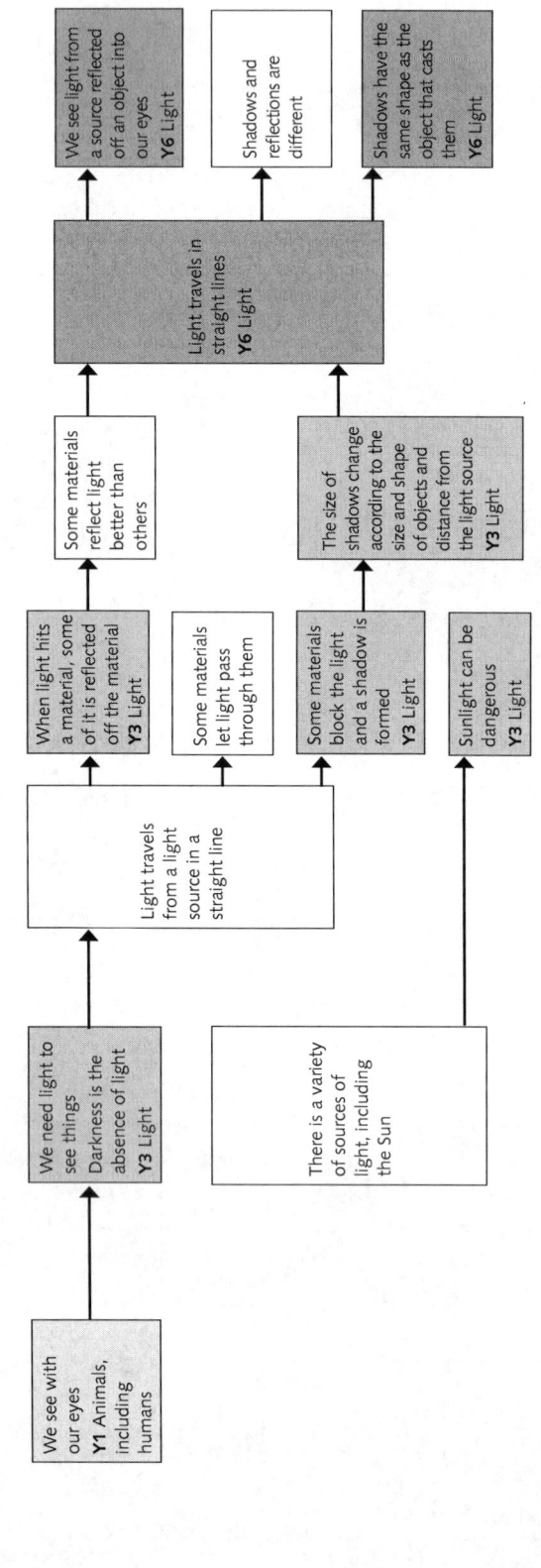

SOUND

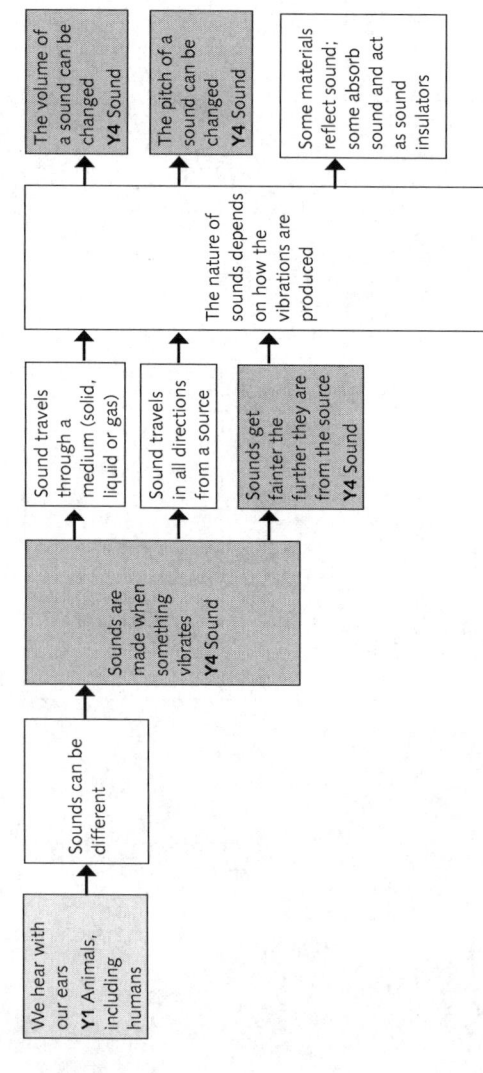

ELECTRICITY

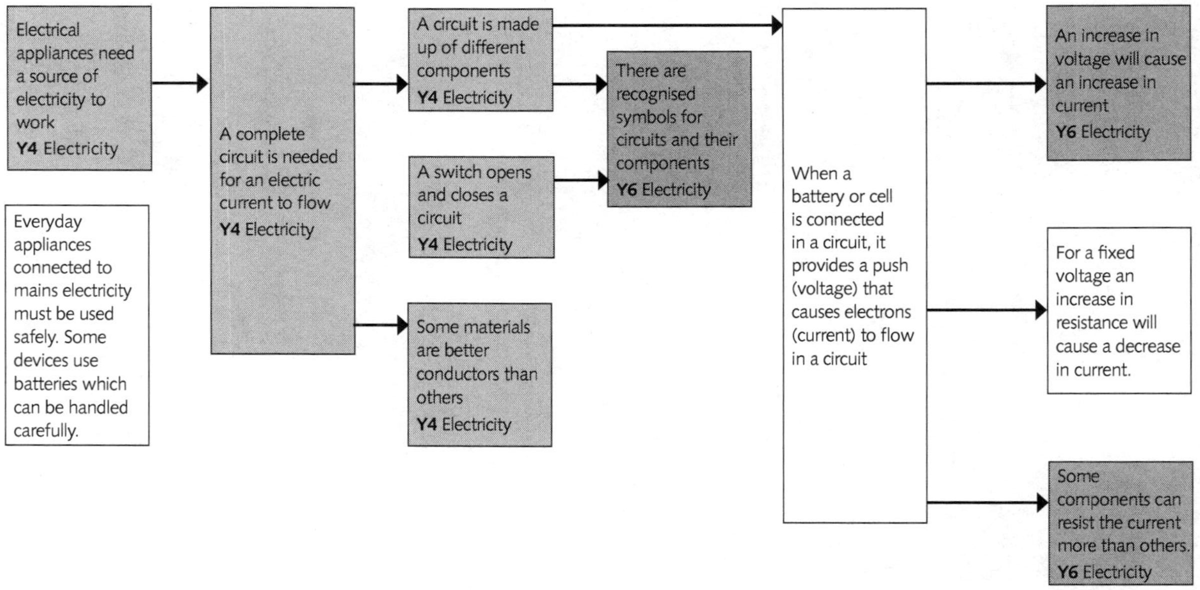

Electrical appliances need a source of electricity to work **Y4** Electricity

A complete circuit is needed for an electric current to flow **Y4** Electricity

A circuit is made up of different components **Y4** Electricity

A switch opens and closes a circuit **Y4** Electricity

Some materials are better conductors than others **Y4** Electricity

There are recognised symbols for circuits and their components **Y6** Electricity

Everyday appliances connected to mains electricity must be used safely. Some devices use batteries which can be handled carefully.

When a battery or cell is connected in a circuit, it provides a push (voltage) that causes electrons (current) to flow in a circuit

An increase in voltage will cause an increase in current **Y6** Electricity

For a fixed voltage an increase in resistance will cause a decrease in current.

Some components can resist the current more than others. **Y6** Electricity

FORCES

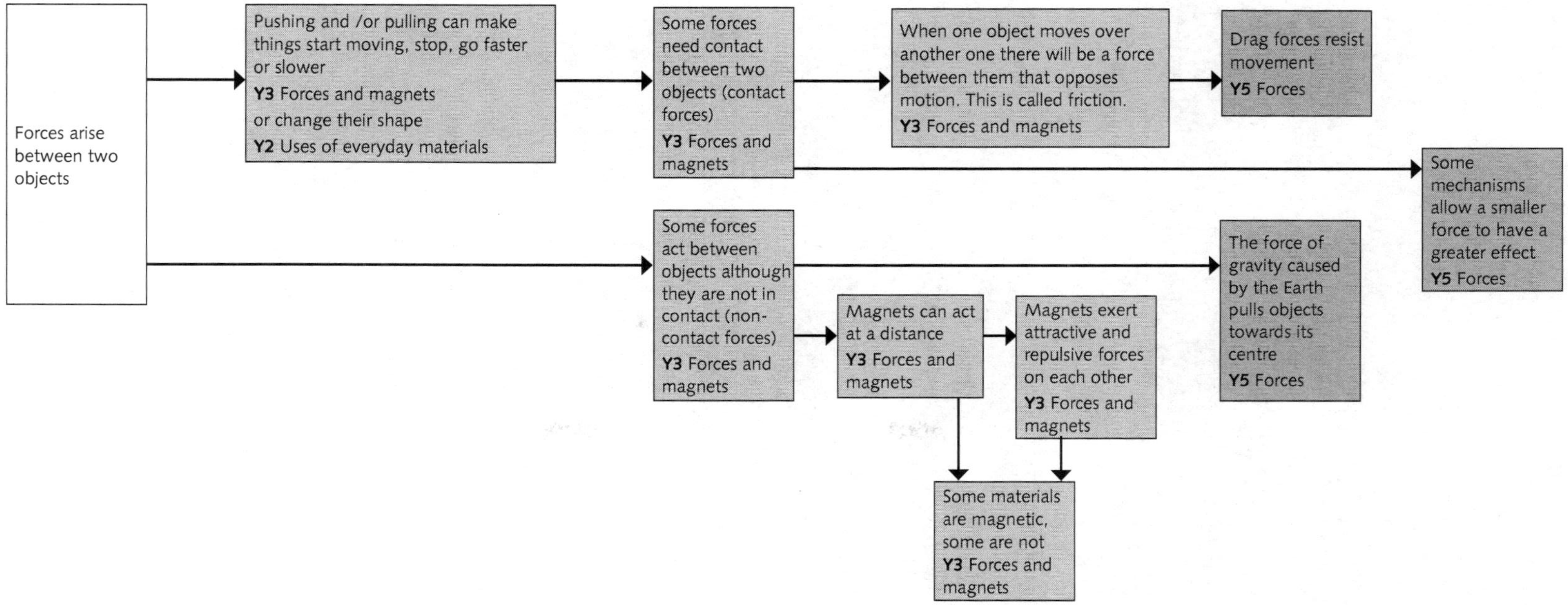

Forces arise between two objects

Pushing and /or pulling can make things start moving, stop, go faster or slower **Y3** Forces and magnets **Y2** Uses of everyday materials

Some forces need contact between two objects (contact forces) **Y3** Forces and magnets

Some forces act between objects although they are not in contact (non-contact forces) **Y3** Forces and magnets

When one object moves over another one there will be a force between them that opposes motion. This is called friction. **Y3** Forces and magnets

Drag forces resist movement **Y5** Forces

Some mechanisms allow a smaller force to have a greater effect **Y5** Forces

Magnets can act at a distance **Y3** Forces and magnets

The force of gravity caused by the Earth pulls objects towards its centre **Y5** Forces

Magnets exert attractive and repulsive forces on each other **Y3** Forces and magnets

Some materials are magnetic, some are not **Y3** Forces and magnets

EARTH IN SPACE

Temperature and day length changes over the year – this pattern is referred to as the seasons **Y1** Seasonal change

→ The Sun appears to move across the sky

→ The Earth, Sun and Moon are approximately spherical **Y5** Earth and space

→ The Earth is one of eight planets that orbit the Sun **Y5** Earth and space

→ The Moon orbits the Earth and looks different at different times of the month **Y5** Earth and space

The Earth orbits the Sun once every year **Y5** Earth and space

→ The seasons change as the Earth's position changes relative to the Sun

The Earth rotates on its own axis once every 24 hours **Y5** Earth and space

→ It is due to the rotation of the earth that we experience day and night **Y5** Earth and space

WORKING SCIENTIFICALLY

Developing independence and autonomy in raising questions, planning and carrying out investigations

Approaches to enquiry

Children should be helped to develop their understanding of scientific ideas by using different types of scientific enquiry to answer their own questions, including:

- observing changes over a period of time
- noticing patterns
- grouping and classifying things
- carrying out simple comparative tests
- finding things out using secondary sources of information

Children should ask their own questions about what they observe and make some decisions about which types of scientific enquiry are likely to be the best ways of answering them including:

- observing changes over time
- noticing patterns
- grouping and classifying things
- carrying out simple fair tests
- finding things out using secondary sources of information

Children should select the most appropriate ways to answer science questions using different types of scientific enquiry, including:

- observing changes over different periods of time
- noticing patterns
- grouping and classifying things
- carrying out fair tests
- finding things out using a wide range of secondary sources of information

Asking questions

Ask simple questions

- Begin to shape questions using different question stems
- Ask questions about how and why objects, materials and living things:
  o change
  o are similar or different to each other
  o connect with each other
  o are made or work
- Suggest questions to investigate

Ask relevant questions

- Recognise questions that can be investigated scientifically and those that cannot
- Ask a clear scientific question
- Recognise when questions can be answered by first hand or second sources of evidence

Use results to raise further questions

- Independently ask questions and offer ideas for scientific enquiry

Use test results to make predictions to set up further comparative and fair tests

Planning

Recognise that questions can be answered in different ways

- With support:
  o Suggest how to find things out
  o Identify changes to observe and measure
  o Identify patterns to observe and measure
  o Identify variables to change and measure
  o Identify sorting criteria
  o Suggest how to take measurements
  o Suggest next steps or a sequence of steps in a plan

Use different types of scientific enquiries to answer them

- Identify different ways to answer a question
- Choose the most appropriate method

Set up simple practical enquiries, comparative and fair tests

- Decide what observations to make, how often and what equipment to use
- Decide what measurements to take, how long to make them for and whether to repeat them
- Decide what sorting or classification criteria to use
- Recognise when a simple fair test is necessary
- With help, decide what variables to change and measure

Plan different types of scientific enquiries to answer questions

- Explain why an enquiry method is the most appropriate to answer a question
- Plan systematic collection of data and which equipment to use
- Plan collection of sufficient data
- Recognise when research using secondary sources will answer questions
- Decide which sources of information to use to answer questions

Recognise and control variables where necessary

- Recognise when variables need to be controlled and why
- Recognise when variables cannot be controlled and a pattern seeking enquiry is appropriate
- Identify which variables have the greatest effect on the result

Becoming more systematic and accurate in collecting, recording and presenting data

Collecting data

Observe closely, using simple equipment
- Choose and use appropriate simple equipment to make observations
- Use non-standard units to collect observations

performing simple tests
- Choose and use appropriate simple equipment with increasing accuracy to collect comparative data
- Use non-standard units to collect data

identifying and classifying
- Sort objects by observable and behavioural features
- Make comparisons between simple features

gathering data to help in answering questions
- Gather data to answer questions from a variety of sources including talking to people, simple books and electronic media, first hand observation and practical activity

Make systematic and careful observations where appropriate, take accurate measurements using standard units, using a range of equipment, including thermometers and data loggers
- Use a range of equipment including data loggers to collect data using standard measures
- With support take accurate measurements on measuring equipment, recognising when to repeat them
- Carry out simple tests to sort and classify materials according to properties or behaviour

Gather data in a variety of ways to help in answering questions
- Gather data to answer questions from a variety of sources including using textbooks, simple keys, electronic media, first hand observation, practical activity and data collected by others

Take measurements, using a range of scientific equipment with increasing accuracy and precision
- Use a range of equipment accurately without support to collect observations and measurements
- Repeat sets of observations or measurements, where appropriate, selecting suitable ranges and intervals
- Use a series of tests to sort and classify materials
- Use relevant information and data from a range of secondary sources to answer questions

Presenting data

Record data to help in answering questions
- Talk about what has been found out and how
- Record observations in word and pictures
- Record observations and test results in simple prepared pictograms, tables, tally charts, bar charts and maps including ICT formats
- Record sorting in sorting circles or tables

Record data in a variety of ways to help in answering questions
- Make notes
- Record data in tables and bar charts
- Use graphs produced by data loggers

Classify in a variety of ways to help in answering questions
- Use Carroll diagrams, and Venn diagrams to classify
- Use and make simple keys to identify and classify

Present data in a variety of ways to help in answering questions
- Drawings, labelled diagrams
- Bar charts, bar line graphs, simple scatter graphs and tables using ICT where appropriate

Record data and results of increasing complexity using scientific diagrams and labels, classification keys, tables and bar and line graphs and models
- Decide how to record data accurately and appropriately
- Use appropriate scientific language in oral and written presentations
- Make keys and branching databases with 4 or more items
- Use more than one source of scientific evidence to identify and classify things
- Present data in line graphs, scatter graphs and frequency charts

Increasingly using scientific knowledge and understanding in conclusions and explanations

Concluding

Use their observations and ideas to suggest answers to questions

- Use simple scientific language to talk about observation or findings
- Use results to answer the investigation question
- Identify simple changes
- Sequence changes
- Say whether the change was expected
- Identify similarities and differences
- Make simple comparisons
- Make links between two sets of observations
- Identify simple patterns and talk about them
- Say whether the pattern was expected
- Identify simple causal relationships
- Say if the relationship was expected

Report on findings from enquiries, including oral and written explanations, displays or presentations of results and conclusions

- Draw simple conclusions about changes observed and link these to scientific ideas
- Refer to a table or graph when reporting findings
- Begin to use and interpret graphs produced by data loggers
- Draw a simple conclusion about similarities and differences identified and link these to scientific ideas
- Draw conclusions about simple patterns between two sets of data
- Draw simple causal conclusions from fair tests
- Draw conclusions from data from different secondary sources

Identify differences, similarities or changes related to simple scientific ideas and processes

- Make links between:
  - o observed changes
  - o similarities and differences
  - o simple patterns between two sets of data
  - o simple causal relationships
  - o data from secondary sources
- and simple scientific ideas and processes

Use straightforward scientific evidence to answer questions or to support their findings

Refer to evidence from practical tests and observations or from secondary data sources when answering questions or explaining findings

- Use simple scientific language in a range of oral and written presentations suitable for different audiences to present findings

Report and present findings from enquiries, including conclusions, causal relationships and explanations of results in written forms such as displays and other presentations

- Use scientific evidence to answer questions or support findings
- Draw valid conclusions about changes, similarities and differences, and causal relationships from data collected
- Draw valid conclusions that utilise more than one piece of supporting evidence
- Use scientific knowledge to explain findings
- Use simple models to help describe scientific ideas
- Explain differences in repeated observations or measurements, identifying reasons for any anomalies noticed

Communicate findings in written form, displays, multi-media and other forms of presentation using scientific language

Evaluating

- Say whether data was useful
- Say whether an information source was useful

Give an opinion about some further information

Use results to draw simple conclusions, make predictions for new values, suggest improvements, and raise further questions

- Make predictions for new values within or beyond the collected data collected
- Identify new questions arising from the data
- Find ways of improving enquiries

Identify scientific evidence that has been used to support or refute ideas or arguments

- Begin to separate opinion from fact
- Use scientific evidence to justify ideas
- Talk about how scientific ideas have developed over time

Identify when further tests and observations might be needed

Evaluate the effectiveness of their working methods, making practical suggestions for improving them

MODULE OVERVIEW CHART

Module number and name	Lesson number and name	National curriculum links	Working scientifically links	Scientific enquiry type	Lesson summary
Year 6 Our Changing World	1: How do animals behave during different times of the year?	Identify how animals and plants are adapted to suit their environment in different ways and that adaptation may lead to evolution	Recording data and results of increasing complexity using scientific diagrams and labels, classification keys, tables, scatter graphs, and bar and line graphs	Grouping and classifying	During these lessons children build and expand on the work in other biology modules undertaken during Years 5 and 6. They visit a variety of different locations around the school grounds and in the wider environment, identifying examples of animals that they observe and describing what the animals are doing at different times of the year.
	2: How can we observe animals when we are not there?	Describe the differences in the life cycles of a mammal, an amphibian, an insect and a bird; identify how animals and plants are adapted to suit their environment in different ways and that adaptation may lead to evolution	Recording data and results of increasing complexity using scientific diagrams and labels, classification keys, tables, scatter graphs, and bar and line graphs	Pattern seeking	During these lesson children make direct and detailed observations of animals at various times of the year by creating and using observation stations, both around school and at a distance, using webcams.
	3: How can we observe the life cycles of specific animals more closely?	Describe the differences in the life cycles of a mammal, an amphibian, an insect and a bird; recognise that living things produce offspring of the same kind, but normally offspring vary and are not identical to their parents	Recording data and results of increasing complexity using scientific diagrams and labels, classification keys, tables, scatter graphs, and bar and line graphs	Grouping and classifying	This lesson complements the work undertaken in Lesson 1. Children study in detail the life cycle of the butterfly, in a controlled environment, making regular observations and identifying and recording changes that take place within the reproductive cycle of butterflies.
	4: How does the number, type and behaviour of birds found around our school change during the year?	Describe the differences in the life cycles of a mammal, an amphibian, an insect and a bird; identify how animals and plants are adapted to suit their environment in different ways and that adaptation may lead to evolution	Reporting and presenting findings from enquiries, including conclusions, causal relationships and explanations of and degree of trust in results, in oral and written forms such as displays and other presentations	Pattern seeking	During these lesson children observe the variety of bird life located around their school and within the local area throughout the year and identify patterns in the data they collect.
	5: What happens to invertebrates during the year?	Describe how living things are classified into broad groups according to common observable characteristics, similarities and differences, including micro-organisms, plants and animals; identify how animals and plants are adapted to suit their environment in different ways and that adaptation may lead to evolution	Reporting and presenting findings from enquiries, including conclusions, causal relationships and explanations of and degree of trust in results, in oral and written forms such as displays and other presentations	Observing changes over different periods of time	Children monitor invertebrates found in the locality of the school during the year, identifying and recording the numbers of different types of invertebrate and their habitats. They conduct some simple investigations into the behaviour of invertebrates and examine ways in which these animals show adaptation to the environments in which they live.
Year 6 Module 1: The Nature Library	1: Can you sort this mess?	Describe how living things are classified into broad groups according to common observable characteristics and based on similarities and differences, including micro-organisms, plants and animals	Recording data and results of increasing complexity using scientific diagrams and labels, classification keys, tables, scatter graphs, and bar and line graphs	Grouping and classifying	In this lesson children build on their knowledge from previous years about how living things can be grouped together in different ways according to the characteristics they have in common; this is classification.
	2: Can you face the garden centre challenge?	Describe how living things are classified into broad groups according to common observable characteristics and based on similarities and differences, including micro-organisms, plants and animals	Recording data and results of increasing complexity using scientific diagrams and labels, classification keys, tables, scatter graphs, and bar and line graphs	Grouping and classifying	In this lesson children decide ways in which to group plants. They apply their classification skills to different types of plants, giving their reasons for the groups and justifying them to others.
	3: How are vertebrates grouped together?	Describe how living things are classified into broad groups according to common observable characteristics and based on similarities and differences, including animals; to give reasons for classifying animals based on specific characteristics	Reporting and presenting findings from enquiries including conclusions, causal relationships and explanations of and degree of trust in results, in oral and written forms such as displays and other presentations	Grouping and classifying	In this lesson children consider the classification of animals. After revising their knowledge of different types of animals from previous years, they investigate in more detail the grouping and classification of vertebrates.

Module number and name	Lesson number and name	National curriculum links	Working scientifically links	Scientific enquiry type	Lesson summary
	4: How are invertebrates grouped together?	Describe how living things are classified into broad groups according to common observable characteristics and based on similarities and differences, including animals; to give reasons for classifying animals based on specific characteristics	Reporting and presenting findings from enquiries, including conclusions, causal relationships and explanations of and degree of trust in results, in oral and written forms such as displays and other presentations	Grouping and classifying	This lesson follows on from Lesson 3 and deliberately adopts the same pattern and activities in order for children to explore the classification of invertebrates.
	5: Where do things fit?	Describe how living things are classified into broad groups according to common observable characteristics and based on similarities and differences, including animals; to give reasons for classifying animals based on specific characteristics	Recording data and results of increasing complexity using scientific diagrams and labels, classification keys, tables, scatter graphs, and bar and line graphs	Grouping and classifying	In this lesson children apply some of the things they have learned in the previous lessons in this module to living things in the school environment. They use their knowledge of classification and their identification skills to create a school log book.
	6: What else is living besides plants and animals?	Describe how living things are classified into broad groups according to common observable characteristics and based on similarities and differences, including animals; to give reasons for classifying animals based on specific characteristics	Identifying scientific evidence that has been used to support or refute ideas or arguments	Grouping and classifying	In this lesson children are introduced to the idea that plants and animals are only two types of living things and that there are three further kingdoms – fungi, bacteria and protista. Together they are often described as micro-organisms.
	7: How can you grow your own micro-organisms?	Describe how living things are classified into broad groups according to common observable characteristics and based on similarities and differences, including micro-organisms, plants and animals	Planning different types of enquiries to answer questions including recognising and controlling variables where necessary	Observing changes over different periods of time	In this lesson children plan and set up an investigation to observe how micro-organisms grow and multiply over time. The results of this investigation need to be recorded over time. Weekly opportunities to observe changes are needed during the enquire stage, ideally over four weeks.
	8: Was it always this way?	Describe how living things are classified into broad groups according to common observable characteristics and based on similarities and differences, including micro-organisms, plants and animals; give reasons for classifying plants and animals based on specific characteristics	Reporting and presenting findings from enquiries, including conclusions, causal relationships and explanations of and degree of trust in results, in oral and written forms such as displays or other presentations; identifying scientific evidence that has been used to support or refute ideas	Finding things out using a wide range of secondary sources of information	In this lesson children explore the history of classification and the scientists involved, including Aristotle and Carl Linnaeus.
	9: What happens when scientists disagree?	Give reasons for classifying plants and animals based on specific characteristics	Identifying scientific evidence that has been used to support or refute ideas	Finding things out using a wide range of secondary sources of information	In this lesson children consider how scientists can have different views on the way things may be classified by using an example of how botanists developed different ways of classifying plants. They then try to reach agreement on their own classification systems for seeds.
	10: What should we call it?	Give reasons for classifying plants and animals based on specific characteristics	Presenting findings from enquiries in oral and written forms such as displays or other presentations	Grouping and classifying	In this lesson children apply their learning from the module in an unfamiliar context. By the end of this lesson children have carried out a focused activity, using evidence and their own knowledge to classify and name an unknown organism.
	EL1: Can you make a nature guidebook for your school?	Describe how living things are classified into broad groups according to common observable characteristics based on similarities and differences; give reasons for classifying plants and animals based on specific characteristics	Reporting and presenting findings from enquiries including conclusions, causal relationships and explanations of and degree of trust in results, in oral and written forms such as displays and other presentations	Grouping and classifying	In this lesson children apply their knowledge from the module to a real life context by looking at living things that they have in their home and their school environment.

Module number and name	Lesson number and name	National curriculum links	Working scientifically links	Scientific enquiry type	Lesson summary
	EL2: What happens when the last one leaves?	Describe how living things are classified into broad groups according to common observable characteristics based on similarities and differences; give reasons for classifying plants and animals based on specific characteristics	Identifying scientific evidence that has been used to support or refute ideas or arguments	Grouping and classifying	In this lesson children apply their classification skills to extinct or nearly extinct things.
Module 2: Body Pump	1: What does my circulatory system do?	Identify and name the main parts of the human circulatory system and describe the functions of the heart, blood vessels and blood	Recording data and results of increasing complexity using scientific diagrams and labels, classification keys, tables, scatter graphs, and bar and line graphs	Finding things out using a wide range of secondary sources of information	In this lesson children begin their investigations of the human circulatory system, first revising knowledge of the digestive, muscular and skeletal systems
	2: What is a heart and what does it do?	Identify and name the main parts of the human circulatory system and describe the functions of the heart, blood vessels and blood	Reporting and presenting findings from enquiries, including conclusions, causal relationships and explanations of and degree of trust in results, in oral and written forms such as displays and other presentations	Finding things out using a wide range of secondary sources of information	In this lesson children make a model of the heart to illustrate how the different parts fit and work together.
	3: What is blood?	Identify and name the main parts of the human circulatory system and describe the functions of the heart, blood vessels and blood	Identifying scientific evidence that has been used to support or refute ideas or arguments	Finding things out using a wide range of secondary sources of information	In this lesson children pose and answer different types of questions to find out how blood transports oxygen and waste gases round the body.
	4: What is in blood?	Identify and name the main parts of the human circulatory system and describe the functions of the heart, blood vessels and blood	Reporting and presenting findings from enquiries, including conclusions, causal relationships and explanations of and degree of trust in results, in oral and written forms such as displays and other presentations	Finding things out using a wide range of secondary sources of information	In this lesson children make 'blood soup' as an illustrative practical activity to help them find out about how the different parts of blood enable it to carry oxygen, waste gases, nutrients and water, and compile a fact file.
	5: What do valves and blood vessels do?	Identify and name the main parts of the human circulatory system and describe the functions of the heart, blood vessels and blood	Recording data and results of increasing complexity using scientific diagrams and labels, classification keys, tables, scatter graphs, and bar and line graphs	Finding things out using a wide range of secondary sources of information	In this lesson children use their learning from previous lessons in this module and secondary sources to explore valves and blood vessels. They create concept sentences and maps to present their findings about valves, veins, arteries and capillaries.
	6: What happens to water in our bodies?	Describe the ways in which nutrients and water are transported within animals, including humans	Reporting and presenting findings from enquiries, including conclusions, causal relationships and explanations of and degree of trust in results, in oral and written forms such as displays and other presentations	Finding things out using a wide range of secondary sources of information	In this lesson children learn more about how water is transported through their bodies, building on their knowledge about how the blood transports nutrients and gases.
	7: What does the road around our body look like?	Identify and name the main parts of the human circulatory system and explain the functions of the heart, blood vessels and blood; to describe the ways in which nutrients and water are transported within animals, including humans	Identifying evidence that has been used to support and refute ideas or arguments	n/a	In this lesson children reflect on what sources of evidence they have used to learn about the human circulatory system and demonstrate their understanding in a card sort and by making a game.

Module number and name	Lesson number and name	National curriculum links	Working scientifically links	Scientific enquiry type	Lesson summary
Module 3: Body Health	1: What does being healthy mean?	Recognise the impact of diet, exercise, drugs and lifestyle on the way bodies function	Reporting and presenting findings from enquiries, including conclusions, causal relationships and explanations of and degree of trust in results, in oral and written forms such as displays and other presentations	Finding things out using a wide range of secondary information	In this lesson children revise their learning about how humans obtain nutrition from the different types of food they eat.
	2: How is food divided into different groups?	Recognise the impact of diet, exercise, drugs and lifestyle on the way bodies function	Identifying scientific evidence that has been used to support or refute ideas or arguments	Grouping and classifying	In this lesson children examine food packaging labels to identify the food groups that different types of food contain, using their existing knowledge of the four main food groups.
	3: What makes a healthy snack or drink?	Recognise the impact of diet, exercise, drugs and lifestyle on the way bodies function	Recording data in a table and reporting and presenting findings from enquiries, including conclusions, causal relationships and explanations of and degree of trust in results, in oral and written forms such as displays and other presentations	Finding things out using a wide range of secondary sources of information	In this lesson children build on their learning from Lesson 2 to examine the nutritional content of certain snacks and drinks to decide whether they would contribute to a balanced, healthy diet. They examine different food packaging and look at how carefully worded food packaging can sometimes be misleading to the purchaser.
	4: How have diets changed?	Recognise the impact of diet, exercise, drugs and lifestyle on the way bodies function	Identifying scientific evidence that has been used to support or refute ideas or arguments	Finding things out using a wide range of secondary sources of information	In this lesson children look at how ideas have been tested scientifically to identify cause and effect and how the results have impacted our diet. They investigate historical cases of diet affecting health, including scurvy and the work of scientist James Lind.
	5: How is pulse rate affected by exercise?	Recognise the impact of diet, exercise, drugs and lifestyle on the way bodies function	Taking measurements, using a range of scientific equipment, with increasing accuracy and precision, taking repeat readings where appropriate; reporting and presenting findings from enquiries, including degree of trust in results	Carrying out comparative and fair tests	In this lesson children explore the impact of exercise on the body. They learn that they can measure their pulse rate to find out how hard their heart is working. They measure their resting heart rate and collect data to investigate what happens when they exercise.
	6: What are the benefits of sports and exercise?	Recognise the impact of diet, exercise, drugs and lifestyle on the way bodies function	Reporting and presenting findings from enquiries, including conclusions, causal relationships, in oral and written forms such as displays and other presentations and explanations of and degree of trust in results	Finding things out using a wide range of secondary sources of information	In this lesson children survey the range of sports played by their classmates, consider the importance of exercise for a healthy lifestyle and develop ways to encourage more people to take up a new sport.
	7: How do drugs affect the body over time?	Recognise the impact of diet, exercise, drugs and lifestyle on the way bodies function	Presenting findings including causal relationships in oral and written forms	Finding things out using a wide range of secondary sources of information	In this lesson children explore the impact of drugs on the way the body functions.
	8: How does smoking affect the body?	Recognise the impact of diet, exercise, drugs and lifestyle on the way bodies function	Reporting and presenting findings from enquiries, including conclusions, causal relationships and explanations of and degree of trust in results, in oral and written forms such as displays and other presentations	Finding things out using a wide range of secondary sources of information	In this lesson children investigate the risks posed to health by smoking. They explore the laws associated with smoking and the short- and long-term health risks associated with smoking.
	9: Can you spread the healthy word?	Recognise the impact of diet, exercise, drugs and lifestyle on the way bodies function	Reporting and presenting findings from enquiries, conclusions, including causal relationships and explanations of and degree of trust in results, in oral and written forms such as displays and other presentations	n/a	In this lesson children produce a school booklet about the benefits of a healthy lifestyle. Children are required to reflect on their learning and present their findings to an audience.

Module number and name	Lesson number and name	National curriculum links	Working scientifically links	Scientific enquiry type	Lesson summary
	EL1: How do athletes keep fit?	Recognise the impact of diet, exercise, drugs and lifestyle on the way bodies function	Reporting and presenting findings from enquires, conclusions, including causal relationships and explanations of and degree of trust in results, in oral and written forms such as displays and other presentations	Finding things out using a wide range of secondary sources of information	In this lesson children look in detail at the diet and training regimes of a range of athletes.
	EL2: What happens when athletes cheat?	Recognise the impact of diet, exercise, drugs and lifestyle on the way bodies function	Reporting and presenting findings from enquiries, including conclusions, causal relationships and explanations of and degree of trust in results, in oral and written forms such as displays and presentations	Finding things out using a wide range of secondary sources of information	In this lesson children explore why and how athletes use performance-enhancing drugs to enhance their performance, looking at some high-profile cases of the use of banned substances.
Module 4: Everything Changes	1: Why do living things vary?	Recognise that living things produce offspring of the same kind, but that offspring normally vary and are not identical to their parents	Recording data and results of increasing complexity using scientific diagrams and labels, classification keys, tables, scatter graphs, and/or bar and line graphs	Grouping and classifying	In this lesson children investigate and discuss how characteristics of living things, for example, height, size or colour, vary from individual to individual.
	2: Can you breed a dog for a specific purpose?	Recognise that living things produce offspring of the same kind, but that offspring normally vary and are not identical to their parents	Identifying scientific evidence that has been used to support or refute ideas or arguments	Finding things out using a wide range of secondary sources of information	In this lesson children develop their understanding of inheritance and explore how characteristics are passed on from parents to offspring. They sort dogs into breeding pairs in order to produce offspring with particular characteristics. Some children extend these ideas to plants.
	3: How can we make our food better?	Recognise that living things produce offspring of the same kind, but that offspring normally vary and are not identical to their parents	Identifying scientific evidence that has been used to support or refute ideas or arguments	Finding things out using a wide range of secondary sources of information	In this lesson children build on the selective breeding activities in Lesson 2 and extend their learning to the subject of selective breeding for food, and its advantages and disadvantages.
	4: How does the environment affect plants?	Identify how animals and plants are adapted to suit their environment in different ways and that adaptation may lead to evolution	Planning different types of scientific enquiries to answer questions, including recognising and controlling variables	Carrying out comparative and fair tests	In this lesson children begin to investigate ways in which the environment can affect how plants grow. They make observations, and plan and set up a fair test to investigate a demonstrable effect that the environment has on plants.
	5: How do environmental variables affect plants?	Identify how animals and plants are adapted to suit their environment in different ways and that adaptation may lead to evolution	Planning different types of scientific enquiries to answer questions, including recognising and controlling variables	Carrying out comparative and fair tests	In this lesson children continue to develop their knowledge and understanding of how environmental variables affect plant populations by carrying out and analysing the results of the investigations they planned in Lesson 4.
	6: How do living things survive?	Identify how animals and plants are adapted to suit their environment in different ways and that adaptation may lead to evolution	Reporting and presenting findings from enquiries, including conclusions, causal relationships and explanations of results, in oral and written forms such as displays and other presentations	Finding things out using a wide range of secondary sources of information	In this lesson children continue to develop their understanding of the idea that changes in the environment can impact on living things. They examine ways in which the physical features and behaviour of living things make them more suited to the particular habit in which they live, and how adaptations of living things help them to survive in their environment.
	7: Why do living things become extinct?	Identify how animals and plants are adapted to suit their environment in different ways and that adaptation may lead to evolution	Identifying scientific evidence that has been used to support or refute ideas or arguments	Finding things out using a wide range of secondary sources of information	In this lesson children apply their knowledge of how changes in an environment can cause living things to become extinct.

Module number and name	Lesson number and name	National curriculum links	Working scientifically links	Scientific enquiry type	Lesson summary
	8: What does it take to survive?	Identify how animals and plants are adapted to suit their environment in different ways and that adaptation may lead to evolution	Reporting and presenting findings from enquiries, including conclusions, causal relationships and explanations of and degree of trust in results, in oral and written forms such as displays and other presentations	n/a	In this lesson children explore further what living things need in order to survive by designing imaginary animals that have adapted to suit a specific environment. It provides a good opportunity for children to reflect on their learning so far in this module about variation and adaptation.
	9: What evidence is there that living things have changed over time?	Recognise that living things have changed over time and that fossils provide information about living things that inhabited the Earth millions of years ago	Identifying scientific evidence that has been used to support or refute ideas or arguments	Finding things out using a wide range of secondary sources of information	In this lesson children use fossils to examine how plants and animals may have looked in the past and, based on their features, suggest the environment in which they may have lived.
	10: How does natural selection work?	Recognise that living things have changed over time and that fossils provide information about living things that inhabited the Earth millions of years ago	Identifying scientific evidence that has been used to support or refute ideas or arguments	Finding things out using a wide range of secondary sources of information	In this lesson children explore how natural selection works.
	EL1: How can one type of animal become two?	Identify how animals and plants are adapted to suit their environment in different ways and that adaptation may lead to evolution	Identifying scientific evidence that has been used to support or refute ideas or arguments	Finding things out using a wide range of secondary sources of information	In this lesson children apply their knowledge of natural selection to the more complex process of speciation.
Module 5: Danger: Low Voltage!	1: How many simple circuits can you make?	Use recognised symbols when representing a simple circuit in a diagram	Recording data and results of increasing complexity using scientific diagrams and labels, classification keys, tables, scatter graphs, bar and line graphs	Carrying out simple comparative and fair tests	In this lesson children revise and build on their work from Year 4 on how to construct simple circuits.
	2: What does a switch do?	Compare the functions of different components, giving reasons for variations in how components function, including the brightness of bulbs, the loudness of buzzers and the on/off positions of switches, and use recognised symbols when representing a simple circuit in a diagram	Recording data and results of increasing complexity using scientific diagrams and labels	Carrying out simple comparative and fair tests	In this lesson children make and control simple circuits using purchased switches and classroom-made switches.
	3: How strong is your resistance?	Associate the brightness of a lamp or the volume of a buzzer with the number and voltage of cells used in the circuit, compare and give reasons for variations in how components function, including the brightness of bulbs, the loudness of buzzers and the on/off position of switches, and use recognised symbols when representing a simple circuit in a diagram	Reporting and presenting findings from enquiries, including conclusions, causal relationships and explanations of and degree of trust in results, in oral and written forms such as displays and other presentations	Carrying out simple comparative and fair tests	In this lesson children add different components to electrical circuits and role play the flow of electrons in a circuit to explain the idea of resistance.

Module number and name	Lesson number and name	National curriculum links	Working scientifically links	Scientific enquiry type	Lesson summary
	4: Do you know your circuit diagrams and can you construct working circuits from them?	Associate the brightness of a lamp or the volume of a buzzer with the number and voltage of cells used in the circuit, compare and give reasons for variations in how components function, including the brightness of bulbs, the loudness of buzzers and the on/off position of switches, and use recognised symbols when representing a simple circuit in a diagram	Recording data and results of increasing complexity using scientific diagrams, classification keys, tables, scatter graphs, bar and line graphs	Carrying out simple comparative and fair tests	In this lesson children consolidate their learning on circuits and recognised electrical symbols from the previous three lessons.
	5: Will the lights stay on? (Part 1)	There are no direct links to the three statements in the science national curriculum, as these two lessons involve carrying out research and constructing reports about electricity in everyday use	Reporting and presenting findings from enquiries in oral and written forms	Finding things out using secondary sources of information	In these two lessons children research how electricity is generated in different ways. During this first part children prepare debates about different methods of electricity generation, transmission and the siting of generating plants, which they present to the class in Lesson 6. They learn to recognise which secondary sources are most useful to research their ideas and to use relevant scientific language in their debates.
	6: Will the lights stay on? (Part 2)	There are no direct links to the three statements in the science national curriculum, as these two lessons involve carrying out research and constructing reports about electricity in everyday use	Identifying scientific evidence that has been used to support or refute ideas or arguments	Finding things out using secondary sources of information	In these two lessons children are researching how electricity is generated in different ways. In the previous lesson children prepared debates about different methods of electricity generation and transmission, and the siting of generating plants.
	EL1: Are you all wired up? (Part 1)	Associate the brightness of a lamp or the volume of a buzzer with the number and voltage of cells used in the circuit, compare and give reasons for variations in how components function, including the brightness of bulbs, the loudness of buzzers and the on/off position of switches, and use recognised symbols when representing a simple circuit in a diagram	Recording data and results of increasing complexity using scientific diagrams and labels, classification keys, tables, scatter graph and/or bar and line graphs	Carrying out simple comparative and fair tests	In this lesson children use their knowledge of electrical circuits and circuit diagrams to construct circuits of different complexity.
	EL2: Are you all wired up? (Part 2)	Associate the brightness of a lamp or the volume of a buzzer with the number and voltage of cells used in the circuit, compare and give reasons for variations in how components function, including the brightness of bulbs, the loudness of buzzers and the on/off position of switches, and use recognised symbols when representing a simple circuit in a diagram	Reporting and presenting findings from enquiries, including conclusions, causal relationships and explanations of and degree of trust in results in oral and written forms such as displays and other presentations	Carrying out simple comparative and fair tests	In this lesson children complete their discussions as design engineers, build technicians and PR departments, using information from Enrichment Lesson 1. In their challenge groups children select and present information about their circuits.
	EL3: Can you protect the crown jewels? (Part 1)	Associate the brightness of a lamp or the volume of a buzzer with the number and voltage of cells used in the circuit, compare and give reasons for variations in how components function, including the brightness of bulbs, the loudness of buzzers and the on/off position of switches, and use recognised symbols when representing a simple circuit in a diagram	Recording data and results of increasing complexity using scientific diagrams and labels, classification keys, tables, scatter graph and/or bar and line graphs	Carrying out simple comparative and fair tests	In this lesson children use their knowledge of circuits to construct burglar alarms.

Module number and name	Lesson number and name	National curriculum links	Working scientifically links	Scientific enquiry type	Lesson summary
	EL4: Can you protect the crown jewels? (Part 2)	Associate the brightness of a lamp or the volume of a buzzer with the number and voltage of cells used in the circuit, compare and give reasons for variations in how components function, including the brightness of bulbs, the loudness of buzzers and the on/off position of switches, and use recognised symbols when representing a simple circuit in a diagram	Reporting and presenting findings from enquires, including conclusions, causal relationships and explanations of and degree of trust in results in oral and written forms such as displays and other presentations	Carrying out simple comparative and fair tests	In this lesson children complete the construction of their burglar alarm circuits and the PR departments present their combined brochure for the burglar alarm to the class.
Module 6: Light Up Your World	1: What is light and what does it do?	Explain that we see things because light travels from light sources to our eyes or from light sources to objects and then to our eyes	Identifying scientific evidence that has been used to support or refute ideas or arguments	n/a	In this lesson children carry out illustrative practical activities to review their knowledge and understanding about the behaviour of light, including light sources and shadows from Year 3.
	2: Can you see more than just your face in a mirror?	Use the idea that light travels in straight lines to explain that objects are seen because they give out or reflect light into the eye	Using test results to make predictions to set up further comparative and fair tests	Noticing patterns	In this lesson children carry out illustrative practical activities to review and develop their knowledge and understanding of how mirrors work from Year 3, Module 3.
	3: Can light go round corners?	Recognise that light appears to travel in straight lines; use the idea that light travels in straight lines to explain that objects are seen because they give out or reflect light into the eye	Recording data and results of increasing complexity using scientific diagrams and labels, classification keys, tables, scatter graphs, and bar and line graphs	n/a	In this lesson children develop their understanding of mirrors from Lesson 2 and use this to develop a model of how light travels.
	4: Can you make a camera with a box, paper and a pin?	Recognise that light appears to travel in straight lines; use the idea that light travels in straight lines to explain that objects are seen because they give out or reflect light into the eye; explain that we see things because light travels from light sources to our eyes or from light sources to objects and then to our eyes	Recording data and results of increasing complexity using scientific diagrams and labels, classification keys, tables, scatter graphs, and bar and line graphs; identifying scientific evidence that has been used to support or refute ideas or arguments	n/a	In this lesson the idea that light travels in straight lines is reinforced through an illustrative practical activity where children investigate how a pinhole camera works
	5: How can you measure a shadow?	Use the idea that light travels in straight lines to explain why shadows have the same shape as the objects that cast them	Planning different types of scientific enquiries to answer questions, including recognising and controlling variables where necessary	Carrying out comparative and fair tests	In this lesson children will build on their learning about shadows from Year 3, and about the movement of the Earth in space in Year 5, to plan fair tests to investigate how different variables affect the size of a shadow.
	6: What do we know about changing shadow sizes?	Use the idea that light travels in straight lines to explain why shadows have the same shape as the objects that cast them	Recording data and results of increasing complexity using scientific diagrams and labels, classification keys, tables, scatter graphs, and bar and line graphs	Carrying out comparative and fair tests	In this lesson children carry out the fair test to investigate shadow size that they planned in Lesson 5.
	7: Can light change direction without a mirror?	Recognise that light appears to travel in straight lines	Recording data and results of increasing complexity using scientific diagrams and labels, classification keys, tables, scatter graphs, and bar and line graphs; using test results to make predictions to set up further comparative and fair tests	Exploration	In this lesson children explore the refraction of light and some of the phenomena it creates.
	8: How many ways can you make a rainbow?	Recognise that light appears to travel in straight lines	Recording data and results of increasing complexity using scientific diagrams and labels, classification keys, tables, scatter graphs, and bar and line graphs	n/a	In this lesson children carry out illustrative practical activities to investigate how rainbows are made, together with other light and colour effects

Module number and name	Lesson number and name	National curriculum links	Working scientifically links	Scientific enquiry type	Lesson summary
	9: How much do you know about light?	Recognise that light appears to travel in straight lines; use the idea that light travels in straight lines to explain that objects are seen because they give out or reflect light into the eye; explain that we see things because light travels from light sources to our eyes or from light sources to objects and then to our eyes; use the idea that light travels in straight lines to explain why shadows have the same shape as the objects that cast them	Reporting and presenting findings from enquiries, including conclusions, causal relationships and explanations of and degree of trust in results, in oral and written forms such as displays and other presentations	n/a	In this lesson children summarise and consolidate all of the work on light done in Year 6, and assess what they have learned.
	EL1: How can you make a good shadow puppet?	Recognise that light appears to travel in straight lines; use the idea that light travels in straight lines to explain why shadows have the same shape as the objects that cast them	Reporting and presenting findings from enquires, including conclusions, causal relationships and explanations of and degree of trust in results in oral and written forms such as displays and other presentations	n/a	In this lesson children plan, test and make puppets for a shadow theatre. Extension lesson 2 that follows has the performance and evaluation of the activity.
	EL2: What makes a good shadow puppet theatre show?	Recognise that light appears to travel in straight lines; use the idea that light travels in straight lines to explain why shadows have the same shape as the objects that cast them	Reporting and presenting findings from enquires, including conclusions, causal relationships and explanations of and degree of trust in results in oral and written forms such as displays and other presentations	Noticing patterns	In this lesson children perform a story using shadow puppets, applying their knowledge about how puppets are made to create different effects.

RESOURCE MATRIX

Year 6 Our Changing World	Resources
1: How do animals behave during different times of the year?	Digital cameras, hand lenses, animal identification guides, binoculars, access to the school's social networking sites, if appropriate
2: How can we observe animals when we are not there?	Digital cameras, webcams, wifi electronic microscopes, iPad technology (including relevant apps), online webcams showing breeding birds and other animals, hand lenses, animal identification charts, binoculars
3: How can we observe the life cycles of specific animals more closely?	Video cameras, digital cameras, wifi electronic microscope, iPad (and appropriate apps), hand lenses, butterfly kits, access to the internet, for further research
4: How does the number, type and behaviour of birds found around our school change during the year?	Binoculars, monoculars, video cameras, digital cameras, iPads with appropriate bird ID app, webcams, access to the internet for children to view bird information websites
5: What happens to invertebrates during the year?	Materials for catching invertebrates and setting traps, digital cameras, video cameras, iPads with invertebrate identification apps, nets, sieves, large white sheets, binocular microscopes, magnifying lenses, secondary sources of information, for further research

Year 6 Module 1: The Nature Library	Resources
1: Can you sort this mess?	Large selection of different types of sweets (toffees, chocolates, marshmallows, peppermint creams, liquorice allsorts – some of the selections produced by various manufacturers would be suitable), marker pens, flip chart paper, hoops of different sizes (alternatively, different lengths of string that can be tied to form loops of different sizes)
2: Can you face the garden centre challenge?	Examples of different types of plants to include at least one moss, one fern, a conifer and a flowering plant, with photographs to increase the variety; sticky notes, mini whiteboards
3: How are vertebrates grouped together?	Internet access, secondary reference resources
4: How are invertebrates grouped together?	Internet access, secondary reference resources
5: Where do things fit?	Recording materials, such as collection pots, paintbrushes (as tickling sticks to encourage small living things into the collection pots for classification), magnifying glasses, identification keys, whiteboards and pens, clipboards and pencils, digital cameras or iPads, a Google map of the school grounds (optional), identification resources from http:// www.naturedetectives. org. uk/download/ index/ and http:// gatekeepersgaffy. co.uk/?page_id=907
6: What else is living besides plants and animals?	Microscopes (ideally with greater than x8 or x10 zoom, possibly digital), mushrooms
7: How can you grow your own micro-organism?	Petri dishes (at least four per group), trays (one per group), fresh sliced white bread, stale white sliced bread, brown bread, granary bread, sealable transparent plastic bags, sticky tape, labels, cameras
8: Was it always this way?	

Year 6 Module 1: The Nature Library Continued	Resources
9: What happens when scientists disagree?	Samples of seeds from familiar fruits and vegetables such as cucumber, apples, oranges, tomatoes, strawberries, grapes, pumpkin, pomegranates, aubergines, peppers, chillies or butternut squash; these should be presented to children once they have been removed from the fruit or vegetable, washed and dried; pictures of the plants the seeds came from, trays or boards to sort on, labels for the classification groups, knives to cut open the seeds, hand lenses, binocular microscopes, digital cameras
10: What should we call it?	
EL1: Can you make a nature guidebook for your school?	Access needed to suitable ICT tools to publish the book
EL2: What happens when the last one leaves?	

Year 6 Module 2: Body Pump	Resources
1: What does my circulatory system do?	Chalk or masking tape, three tabard style sports bibs, bicycle or foot pump, stethoscopes or cardboard tubes, sheets of red and blue paper stuck back to back
2: What is a heart and what does it do?	Different coloured modelling clay (enough for children to work in pairs), scissors, base boards, digital camera(s) and cocktail sticks (optional), access to secondary sources of information about the heart, children's labelled diagrams of circulatory system from Lesson 1
3: What is blood?	A large bucket capable of holding nine pints (or five litres) of red liquid such as food colouring in water, five empty one-litre drink cartons, plastic funnel and jug
4: What is in blood?	Plasma – yellowy liquid such as weak orange squash (one cup per soup); red blood cells – lots of small red jelly sweets or chopped up pieces of a raspberry jelly cube; white blood cells – a few small white marshmallows; platelets – small amounts of white rice; one sealable plastic bag per person or group making blood soup, children's diagrams from Lesson 1
5: What do valves and blood vessels do?	Access to secondary sources such as reference books and the internet, scissors and glue
6: What happens to water in our bodies?	Access to secondary sources of information about human and other circulatory systems, such as the internet, books, posters or leaflets
7: What does the road around our body look like?	Scissors, split pins, modelling clay, access to all materials produced in Lessons 1–6, plus secondary sources of information that children have used previously in the module

Year 6 Module 3: Body Health	Resources
1: What does being healthy mean?	Large sheets of flip chart paper, sheets of A3 paper, pens, glue, scissors, plain A4 paper, access to secondary sources, including the internet and healthy education pamphlets and posters, for all levels of challenge
2: How is food divided into different groups?	Range of food packaging either sealed or clean and empty, mini whiteboards, small paper plate (one per child)
3: What makes a healthy snack or drink?	Sticky notes (five per child), wide range of snack food packaging with nutritional value information accessible
4: How have diets changed?	Access to secondary sources for all levels of challenge
5: How is pulse rate affected by exercise?	Stopwatch (one per pair)
6: What are the benefits of sports and exercise?	Access to secondary sources, including the internet, for all levels of challenge: sticky notes
7: How do drugs affect the body over time?	Existing drugs resources your school may already have, access to secondary sources for all levels of challenge
8: How does smoking affect the body?	Two large PE hoops; access to secondary sources about smoking risks, should children wish to use them
9: Can you spread the healthy word?	Access to work produced in previous lessons. Make available a range of presentation tools for children to select from six sheets of A3 plain white paper to collate research; potential to email the head teacher for the Reflect and review session is preferred
EL1: How do athletes keep fit?	
EL2: What happens when athletes cheat?	Access to secondary sources, including the internet, for further research

Year 6 Module 4: Everything Changes	Resources
1: Why do living things vary?	Rulers, metre sticks or tape measures, sticky notes, large sheets of paper (A3), access to the internet or books for further research
2: Can you breed a dog for a specific purpose?	Secondary sources of information for further research
3: How can we make our food better?	Secondary sources of information for further research
4: How does the environment affect plants?	Access to wild plants in different habitats or photographs of wild plants in different habitats
5: How do environmental variables affect plants?	Petri dishes, cotton wool, cress and mustard seeds, dark paper
6: How do living things survive?	Large pieces of paper, access to secondary sources of information, including the internet or books, for further research
7: Why do living things become extinct?	Sheets of A3 paper, secondary sources of information, including the internet, for further research
8: What does it take to survive?	Model-building materials, if available
9: What evidence is there that living things have changed over time?	Collection of fossils, including a fish fossil if possible, or a selection of photographs of fossils, access to secondary sources of information, including the internet, for further research
10: How does natural selection work?	Large pieces of paper, plastic cups, rice, tweezers, tongs, plastic forks, plastic knives, large marbles (if possible)
EL1: How can one type of animal become two?	Access to secondary sources of information, for further research if there is time

Year 6 Module 5: Danger! Low Voltage	Resources
1: How many simple circuits can you make?	Commercially produced energy stick or human circuit ball (available from primary science equipment suppliers, for example, TTS), 1.5 V cells, lamps, lamp holders wire, thin tinfoil strips, cell holders, (one of each component between two), magnifiers, digital magnifier, modelling clay (which is useful to anchor a cell while the circuits are constructed), extra wire, small screwdrivers, mini whiteboards
2: What does a switch do?	A2 paper, 1.5 V cells, lamps, wire, crocodile clips, toggle switches, slide switches, push switches, lamp holders, cell holders, small screwdrivers, wire strippers, match boxes, metal foil, paper fasteners, paper clips, film canisters, small ball bearings, card, adhesive tape or glue, hand drills, drill bit, examples of mains switches
3: How strong is your resistance?	1.5 V and 4.5 V cells, lamps, wire, crocodile clips, switches, lamp holders, cell holders, small screwdrivers, wire strippers, pencil or propelling pencil leads stuck to lollipop sticks, resistance wire and/or different thicknesses of fuse wire
4: Do you know your circuit diagrams and can you construct working circuits from them?	1.5 V cells, lamps, wire, crocodile clips, switches, lamp holders, cell holders, small screwdrivers, wire strippers
5: Will the lights stay on? (Part 1)	Sticky notes, A2 sheets of paper, computers, access to resources to allow children to research electricity generation and using renewable sources of electricity generation
6: Will the lights stay on? (Part 2)	Sticky notes, A2 sheets of paper, ICT hardware for presentations as required
EL1: Are you all wired up? (Part 1)	Cells (1.5 V and 4.5 V), lamps (1.5 V and 2.5 V), yellow green red LEDs, flashing lamps (3 V), wire, crocodile clips, switches, lamp holders, cell holders, small screwdrivers, wire strippers, wire connection block, corrugated plastic sheet or thick card, paper drill or hole punch, paper fasteners, paper clips, scissors, PVA glue, low-melt glue guns, glue gun stands, digital camera/video camera
EL2: Are you all wired up? (Part 2)	Completed circuits from Enrichment Lesson 1, presentation resources as selected by children, for example, projection hardware, recognised symbols cards
EL3: Can you protect the crown jewels? (Part 1)	Cells (1.5 V, 4.5 V), lamps (1.5 V, 2.5 V, 3.5 V), flashing lamps (3 V), buzzers, wire, crocodile clips, switches from Lesson 2, lamp holders, cell holders, small screwdrivers, wire strippers, wire connection block, corrugated plastic sheet, cardboard box, paper drill or hole punch, paper fasteners, paper clips, metal foil, scissors, PVA glue, low-melt glue guns, glue gun stands, glass beads and/ or plastic tiara, digital camera
EL4: Can you protect the crown jewels? (Part 2)	Completed circuits from Enrichment Lesson 3, presentation resources, for example, projection hardware, as selected by children for the brochure

Year 6 Module 6: Light Up Your World	Resources
1: What is light and what does it do?	Torches, sunglasses, mirrors, a collection of materials that are transparent, translucent and opaque, a light meter/data logger
2: Can you see more than just your face in a mirror?	Plastic mirrors (one of each for each group), shiny metal spoons (ideally larger than teaspoons so a reflection of a child's face is clearly visible in them)
3: Can light go round corners?	Plastic mirrors (two for each group), a selection of torches and a small object such as a car or plastic figure (one of each for each group), cardboard, scissors and Sellotape
4: Can you make a camera with a box, paper and a pin?	Plastic mirror, a bright torch, small shoe box, tracing paper, black paper or card (or kitchen foil), scissors and sticky tape, needle or drawing pin; two very large pieces of card, ideally greater than 1m x 1m, one with a triangle cut out of it (side length about 50cm each side) and the other with a smaller circular hole about the size of a tennis ball cut in it; long straight piece of wooden dowel (2m) or ball of thread, large piece of paper (bigger than A3) and thick marker pen
5: How can you measure a shadow?	Torches, large sheets of white paper, tape measures or metre rulers, card to make shapes, scissors, graph paper, fair test planning board
6: What do we know about changing shadow sizes?	Torches, large sheets of white paper, tape measures or rulers, card to make shapes, scissors, graph paper
7: Can light change direction without a mirror?	For demo: Glass beaker or clear 1pt glass, water, cooking oil, pencil, mini whiteboards (if available) Water magnifier: Clear piece of plastic at least 5cm × 5cm with tape around edges, sheet of newspaper, water and dropper if possible, magnifying glass, glass bead The Surprising Coin: 1p or 5p coin, mug, water Oil, water and a pencil: A jam jar or similar sized clear, straight edged plastic container mainly filled with water but with a layer of at least a few cm of cooking oil on the top, pencil Amazing arrows: Jam jar or clear straight edged glass filled with water. Piece of paper with three parallel arrows drawn on it (see Resource sheet 1 and 2, Refraction circus instructions and observation sheets, for more details) Because of the liquids involved, plenty of paper towels in case of any spillages.
8: How many ways can you make a rainbow?	Torch (ideally with as white a beam as possible, such as a bike light), a red torch or fairy lights (optional). For each colour wheel: marble (a little bigger than the hole in middle of the CD), glue, CD, white paper, scissors, coloured pencils (red, orange, yellow, green, blue and violet) or printed colour template. (Alternative version without marble will need card/scissors/ pens and a strong thread.) Bubble blower and mixture. Tray of water, plastic mirror and torch for each group
9: How much do you know about light?	Access to the internet or a range of books on light, A3/A2 poster paper and pens
EL1: How can you make a good shadow puppet?	For each of the groups a torch or light source, a piece of tracing or greaseproof paper (ideally A3 in size) in a cardboard frame for stability. These are the practice screens for each group. To make the characters, thick card, scissors and split pins. Thin wooden kebab sticks (with ends cut off and made safe) or similar to attach the figures to Health and safety
EL2: What makes a good shadow puppet theatre show?	The finished shadow puppets from the previous lesson; a large thin white sheet with a large light source behind it for the screen, suspended off the floor so that the children can get behind it to be able to make their shadows without making a shadow themselves

TEACHING FRAMEWORK

OUR CHANGING WORLD

INTRODUCTION

In this module children build on and apply their knowledge of living things and how they are adapted to particular environments. In particular thcy should draw on studies carried out in Year 5 and Year 6 modules (for example, Circle of Life, The Nature Library and Everything Changes), in order to investigate in more detail the ways in which animal populations are suited to the environments in which they live. This module includes opportunities to study ways in which physical characteristics, patterns of behaviour and life cycles help to adapt organisms and improve their chances of survival. Children use a range of techniques to help them to observe and monitor changes in the environment, the size of populations, and the behaviour of different groups of animals.

Through working scientifically children understand the need to plan all investigations carefully and refine their skills of observation, measurement, testing, recording and communication. Throughout the module they have several opportunities for peer assessment and feedback on their work and that of others.

Key vocabulary:

mammal, amphibian, insect, bird, metamorphosis, tadpole, nymph, pupae, chrysalis, caterpillar, migrate, hibernate, courtship, plumage, habitat, adaptation, behaviour, young, chick, life cycle, egg, pupae, adult, butterfly, nectar, death rate, nest, brood, fledgling, juvenile, diet, migration, resident, invertebrate, mollusc, worm, snail, woodlouse, centipede, millipede, beetle, aphid, adaptation, predator, prey, survival, habitat, question, investigation, fair test, change, measure, predict, prediction, explanation, observations, draw conclusions, justify, analyse

FACT FILE:

It is important to recognise that living things are suited to their environments. This suitability is in part because of their physical adaptations, and also because of their behaviour patterns and life cycles. For example, birds have beaks that are adapted to particular food types, but they may also feed at particular times and in different ways (for example, on the ground or hanging from tree branches).

The behaviours of living things are important because they help to reduce the level of competition with other organisms. Behavioural patterns are more easily seen in animals and some of these are the focus of this module. For example, by feeding on a range of foods animals can reduce the competition for food and may be able to survive when one type of food is scarce. Similarly, the ways in which animals reduce the risk of being eaten by predators include both physical and behavioural adaptations. (for example, camouflage, being nocturnal or moving very fast).

The interactions between organisms also contribute to their survival. In some cases the interaction is beneficial (for example, by providing a habitat or source of food) to one or both of the organisms, but in other situations the interactions are harmful (for example, predation and competition).

The scientific understanding of changes in the populations of living things and their behaviours has improved with the developments of technology, so it is now possible to observe and monitor individuals and populations more closely and over longer periods of time. Improved understanding of how and why organisms live as they do has also come from combining well-planned observational and monitoring studies with controlled experiments and tests.

Common misconceptions:

Children may not realise that the behaviour of organisms is one of the ways in which they are adapted to the environment.

Children may not see the patterns in life cycles and the ways in which organisms are distributed in different environments.

Children may not recognise the interactions between organisms.

Big Cat book links

Life Cycles Sally Morgan 978-0-00-733640-1 Band 16 Sapphire	Explore the fascinating life cycle of the salmon in this highly photographic book

OUR CHANGING WORLD

Key vocabulary:

mammal, amphibian, insect, bird, metamorphosis, tadpole, nymph, pupae, chrysalis, caterpillar, migrate, hibernate, courtship, plumage, habitat, adaptation, behaviour

Resources:

Digital cameras, hand lenses, animal identification guides, binoculars, access to the school's social networking sites, if appropriate

Health and safety:

All children and adults should wash hands after handling plant or animal materials.

Key information:

The degree of evidence of animal life cycles varies, depending on the season. See Resource sheet 1 for details.

 ## LESSON 1: HOW DO ANIMALS BEHAVE AT DIFFERENT TIMES DURING THE YEAR?

LESSON SUMMARY

During these lessons children build and expand on the work in other biology modules undertaken during Years 5 and 6. They visit a variety of different locations around the school grounds and in the wider environment, identifying examples of animals that they observe and describing what the animals are doing at different times of the year. By the end of these lessons children are able to describe examples of animal behaviour and, based on their observations, suggest reasons for the behaviours and relate them to, for example, the stage of the animal's reproductive cycle, its feeding habits, and adaptations that make the animal suited to the environment in which it lives.

Preparation required:

Identify suitable locations for children to explore and develop your own knowledge of what they might see. Suitable locations might include an area of meadow or flower border, a vegetable garden, a bush or patch of bushes, a wall or unkempt verge, large trees or a wooded area and a pond or river. Your local wildlife trust or a city farm may well be able to help if provision around the school is significantly limited. If possible, produce a PowerPoint or photo-story sequence, featuring images of a variety of habitats around your school and within the local environment at different times of the year, to use with children as you introduce this module.

National curriculum links:

Identify how animals and plants are adapted to suit their environment in different ways and that adaptation may lead to evolution

Working scientifically links:

Recording data and results of increasing complexity using scientific diagrams and labels, classification keys, tables, scatter graphs, and bar and line graphs

Learning intention:

To identify, observe and give examples of animal behaviour throughout the year within the local environment

Scientific enquiry type:

Grouping and classifying

Success criteria:

- I can make detailed observations of animals around the school grounds and within the local environment.
- I can describe the stage that an animal has reached in its life cycle.
- I can identify evidence of behavioural changes in animals over time.
- I can suggest reasons for the behaviours that I have observed and relate them to the animal's lifestyle.

EXPLORE:

If possible, show children a photo story or PowerPoint presentation featuring images of a variety of habitats around your school and within the local environment at different times of the year. Children may have studied their school or environment in these contexts before (particularly in Year 4) and they should be encouraged to look back on this prior learning as the basis for these lessons.

Ask: *How do these habitats change during the year? What sort of animals – mammals, amphibians, insects, birds – might you find in each of these locations now? What stages in their life cycles might they have reached now that it is summer/spring/autumn/winter? What behaviours might you observe? Does this change during the year?*

Children's responses here are dependent upon which time of year the lesson takes place (see Key information). Encourage children to think back to their learning from work that they carried out in Year 5, Circle of Life and Reproduction in Plants and Animals, and in Year 6, The Nature Library and Everything Changes.

Discuss with children some ways in which they might record and communicate their findings throughout the year. These may include: creating a wall display on which images and information about children's initial ideas and subsequent evidence from observations might be shared; keeping a Changing World diary; writing a blog for the school website (and/or use of a form of social media to communicate with each other about what they have observed).

Health and safety:

The use of social media must be kept in line with school policy.

ENQUIRE:

Organise children into small groups and, where necessary, identify a specific location/s for them to visit. Some ideas are listed at the start of this lesson plan. Give the groups equipment, such as binoculars and hand lenses, to help them make their observations.

Prompt children to think about how and where to search for evidence, and explain to them that they should look in a wide range of places, including cracks in walls, under large stones or logs and in leaf litter. Remind them that they need to keep still and be quiet when making observations – watching animals takes time and patience. Remind children that they should always put back any stones or logs that they move when they are looking for evidence of animals.

The challenges are differentiated by the level of support provided to children and the questions they are asked to consider.

Challenge 1: Children find evidence of animals in a local environment and record their observations

Provide each group of children with an Animal hunt record sheet to complete (Resource sheet 2). Ask them to look for six different animals listed on the record sheet and to make notes about each animal (the children could use identification cards either during their visits or when they have returned to the classroom), record what they observe and note what life cycle stage they think the evidence suggests the animal has reached.

When they have returned to the classroom, encourage the children to reflect on their observations.

Ask: *What evidence of animals did you see? Could you identify the animals based on the evidence you observed? What did the evidence tell you about the stage/s that the animals had reached in their life cycles? Did you always see the animal? If not, how did you know that it was/had been there?*

Challenge 2: Children find evidence of different types of animals and record their detailed observations and conclusions in a Changing World diary

Explain to the children that they need to find evidence of at least six different animals, including at least one insect, one bird, one mammal and one amphibian. Ask them to record their observations in a Changing World diary and include what they observe, what the animal is doing and what life cycle stage they think the evidence suggests the animal has reached.

When they are back in the classroom, ask the children to reflect on their observations.

Ask: *What evidence of animals did you see? What did the evidence tell you about the stage that the animals have reached in their life cycles? Did you find evidence for all of the animals you were trying to find? If not, what were you looking for and where did you look? Might you find evidence more easily at another time of the year? Why do you think they were behaving as they were (for example, were they feeding, moving around to collect things, 'running away')?*

Challenge 3: Children predict what evidence of animals they will find, anticipate where to find them and gather detailed evidence of different types of animals to record in a Changing World diary

Explain to the children that they need to use their knowledge and understanding of the animals to predict what evidence they might find (given the season), to anticipate locations that would be best to visit and to identify what they need to look for while they are there. Let them know that they need to gather detailed evidence of at least two animals of each type (insects, birds, mammals and amphibians) and that they should record their observations in a Changing World diary.

When they are back in the classroom, ask them to reflect on their observations.

Ask: *Did you find the evidence of the animals you had predicted you would find? Which types of animals were the most difficult to locate? Did you find anything unexpected? What evidence of animal life cycles did you notice? Which stage was the most evident? What were the animals doing? In what ways do you think that their behaviour helped them to survive?*

REFLECT AND REVIEW:

Ask children to reflect on the evidence that they have collected about the animals they found and on the behaviours that the animals were exhibiting. Encourage them to share what they have discovered – particularly anything unexpected. Add their evidence, in the form of photos and sticky notes, to the wall display, if one is being made.

Ask: *What season is it now? What temperature and weather conditions have we experienced recently? Have the weather and temperature affected the evidence of animal life cycles that you were able to find? Do you think that the environmental conditions affected how the animals behaved? If so how, and why? In what ways do you think that the animals' form and behaviour make them suited to the environment in which they live?*

Remind children that they should be thinking about the kinds of observations they need to make over the course of a year, so that each time their observations provide more evidence to help them answer the more different questions. Explain that it is the quality of observations that are important, not just the quantity.

EVIDENCE OF LEARNING:

Listen to the children's responses during the course of the lesson. Are their observations of animals sufficiently detailed? Do they use their prior knowledge (and available identification information) to accurately name the animals they observe? Can they describe the life cycle stage that most animals have reached, relating this to the season and weather conditions, where appropriate? Can they give examples of animal behaviours and relate those behaviours to the lifestyle of the animal. Can children explain how the form and behaviour of an animal might make it suited to its environment?

OUR CHANGING WORLD

Key vocabulary:

Key vocabulary:

mammal, amphibian, insect, bird, metamorphosis, young, chick, plumage, habitat

Resources:

Digital cameras, webcams, wifi electronic microscopes, iPad technology (including relevant apps), online webcams showing breeding birds and other animals, hand lenses, animal identification charts, binoculars

C

LESSON 2: HOW CAN WE OBSERVE ANIMALS WHEN WE ARE NOT THERE?

LESSON SUMMARY:

During these lesson children make direct and detailed observations of animals at various times of the year by creating and using observation stations, both around school and at a distance, using webcams. If setting up webcams is not feasible, an alternative option is to access webcam footage on specialist websites.

By the end of these lessons children are able to identify and describe specific behaviours of animals at different times of year, make comparisons between animals and suggest reasons for any similarities and differences that they have noticed.

Preparation required:

Plan for and prepare locations where it is possible to safely locate animal stations such as nest boxes, hedgehog houses and bug hotels. If possible, provide 'nest box cameras' or similar, to enable children to view at a distance the behaviour of birds and other animals. It is also possible to view webcams online, including those set up annually by specialist providers such as the Hawk and Owl Trust, local wildlife trusts, Woodland Trust, Butterfly Trust, RSPB, BBC Wildlife, BBC Springwatch and BBC Autumnwatch. Accessing these over time will provide children with an invaluable insight into the behaviours and reproductive cycles of animals. Many of these organisations also provide highlights of the webcam recordings so that it is possible to review major events that have taken place previously. Some useful teacher resources can be found at the following web addresses: http://downloads.bbc.co.uk/wildaboutyourgarden/images/get_wild_about_your_garden_pocketguide.pdf (This is information for teachers on encouraging wildlife into the garden.)

http://www.rspb.org.uk/ourwork/teaching/schoolgrounds/breathingplacesschools/advice.aspx (This has current news and issues for teachers to explore with children.)

Ask children to work together in groups to identify and create their own observation stations to place within the school grounds, prior to the first lesson. These might include designing and constructing specific 'habitats'. Some examples (Toad abode, Minibeast mansion, Hedgehog home, Butterfly feeder, Bumblebee home) can be found on the BBC Breathing Spaces website (http://www.bbc.co.uk/breathingplaces/downloads/how_to_poster/). Ideally, some of these observation stations could be remotely viewed through webcam technology. Alternatively, 'hides' might be placed where children could make observations without disturbing visiting animals.

National curriculum links:

Describe the differences in the life cycles of a mammal, an amphibian, an insect and a bird; identify how animals and plants are adapted to suit their environment in different ways and that adaptation may lead to evolution

Working scientifically links:

Recording data and results of increasing complexity using scientific diagrams and labels, classification keys, tables, scatter graphs, and bar and line graphs

Learning intention:

To make detailed observations of animal behaviour, making comparisons and suggesting reasons for the similarities and differences observed

Scientific enquiry type:

Pattern seeking

Success criteria:

- I can make detailed observations of animal behaviour, accessing evidence from a variety of contexts.

- I can compare evidence gathered in different ways, identifying and explaining any gaps I notice and deciding which sources of evidence tend to be most useful.

- I can suggest reasons for any similarities and differences that I have observed between animals.

EXPLORE:

Ask: *How can we observe what animals are doing when we are not there?*

Discuss with children some approaches to observing animals from a distance. Explore some possibilities, for example, organising a rota of people to use webcams. Establish with children the importance of being systematic in making observations, including ensuring that observations are made at the same specific locations or observation stations.

Introduce children to a number of web-based remote cameras that they can plan to follow during the year and in particular during the main breeding season (that is, from May to June or July, depending on the type of animal being observed).

Ask: *What might you observe by using a remote camera to look at an animal? Why is a remote camera a particularly effective way of watching animals? What additional information might you be able to pick up, and why?*

Ensure that children consider the value of not disturbing animals and of being able to observe carefully and regularly over time, regardless of weather and throughout the daylight hours from daybreak to sunset.

ENQUIRE:

Ask children to decide which animal(s) they are going to observe and to develop a plan for when (how often), where (which observation stations) and how (first-hand by establishing a rota and/or using webcams). Emphasise to children that this is a challenging task, but that it is important that they plan all scientific work carefully and clearly.

Explain to children that during these lessons they are going to visit their observation stations to look for evidence of animal activity and record their evidence using appropriate technology, making notes and drawing their observations. Organise children in pairs or small groups and remind them of how they should behave if they are to be able to observe animals effectively. This includes speaking in a whisper (and only if necessary), treading carefully without making a noise and moving slowly. They should stay alert, listening as well as watching.

The challenges are differentiated by the level of detail in the observations to be made and the interpretation of the findings. Children can use a combination of approaches to gather evidence, including direct observation, digital video cameras and webcams. The actual selection depends on what children can successfully manage.

Challenge 1: Children use a digital camera or video camera to record evidence of animals, and complete their Changing World diaries

Ensure that the children have carefully recorded their plans and have decided on which animal/s they are going to observe.

Ask: *Why did you decide to observe this animal/these animals? What do you want to find out? What will you need to observe? How often will you make your observations?*

Ask children to review their observations after they have made multiple visits.

Ask: *Which animals did you see most of? What were they doing? Was there more or less activity than last time you made observations? Why might that have been?*

Challenge 2: Children record evidence of animals for their Changing World diaries, and consider reasons for animal behaviour and how they can improve their observations

Ask the children to complete their Changing World diaries and to organise records (photographs, videos etc.) of their observations and review them. Prompt them to also look back at their observations from any previous visits.

Ask: *What did you observe last time? Which animals did you see most of? What were they doing? Was there more or less activity than last time you made observations? Why might that have been? When do you think you would see most activity? How might you improve the observations that you make? Can you suggest reasons for why an animal is behaving the way it does?*

Challenge 3: Children use a webcam to make detailed observations of animals and their behaviours

Explain to the children that a more detailed set of observations can be made using a webcam watch. Ask them to plan in as much detail as they can about what they are trying to find out and what they need to record to provide the necessary evidence. You may want to give the children the Observation record sheet (Resource sheet 1) as a framework for recording their observations over time. Remind them that observations can include measurements and counts of how often an animal completes a particular behaviour, for example, how long the bird spends on the nest.

Many webcam websites follow an animal or pair of animals with young. If this is what the children are going to be observing in their challenge, prompt them to think about what they should record.

Ask: *How many young are there? What do they look like? What stage of development have they reached? Were they fed by parents while you were watching? What food do they seem to be eating? How much and how often? How much longer before they are fully grown?*

Depending on the animal being observed (and the website), the children might explore their ideas further, researching more background detail to complete the picture. Some webcam websites also have data from previous years that the children may use to compare with their own observations.

REFLECT AND REVIEW:

Encourage children to share some of their observations and their interpretation of them. Encourage children to compare their observations on animals' behaviours with the observations of their peers.

Ask: *What sorts of animals did you notice around the observation station? Why do you think they live there? What were the most common behaviours shown by the animals? Did you see any interactions between the animals, for example, did one chase another away?*

Move on to discuss in more detail with children the quality of their evidence.

Ask: *What evidence were you able to gather at the observation stations and through the webcams? Did you make any measurements or counts? Do you think it would be useful to do so? How would you present your results? How could you make your evidence better? What were the advantages and disadvantages of the different methods for observing the animals?*

Ask children to talk to their partners and to come up with top tips for animal observers (for example, how should the human observers behave?). Take feedback from children and rank their ideas.

EVIDENCE OF LEARNING:

Listen to children's responses throughout the lesson. Do they make sufficiently detailed observations of animals, whether first-hand or as they observe them using a webcam? Can children compare evidence gathered in different ways, suggesting which sources of evidence tend to be more useful than others? Can they explain the advantages (and disadvantages) of the different methods of observation? Do they suggest reasons for any similarities and differences observed, referencing the type of animal, its habitat requirements, weather conditions or time of year, as appropriate?

OUR CHANGING WORLD

Ⓒ

LESSON 3: HOW CAN WE OBSERVE THE LIFE CYCLES OF SPECIFIC ANIMALS MORE CLOSELY?

LESSON SUMMARY

This lesson complements the work undertaken in Lesson 1. Children study in detail the life cycle of the butterfly, in a controlled environment, making regular observations and identifying and recording changes that take place within the reproductive cycle of butterflies.

By the end of these lesson children have observed metamorphosis as it occurs in butterflies. They have identified the length of time spent at every stage of the life cycle and noticed the extent of the physical changes that occur as a result of metamorphosis.

Preparation required:

For these activities you need to purchase butterfly kits. Ensure that the butterfly kits purchased provide sufficient space to house the butterflies, allowing easy access for children to make observations. The species of butterflies raised should be appropriate for the local habitat, so that the butterflies can ultimately be released into the wild. Butterfly kits include a voucher to send away in advance of the time you want to receive live caterpillars (from March to September). These are provided in a growing medium to ensure that they have the right food requirements. Caterpillars of the painted lady butterfly are often used because they are native to the UK, the caterpillars successfully metamorphose, and the butterflies can be released outside school after the adults have been observed for a few days. The life cycle takes 3–5 weeks to complete.

Key vocabulary:

life cycle, metamorphosis, egg, caterpillar, chrysalis, pupae, adult butterfly, nectar, predators, death rate

Resources:

Video cameras, digital cameras, wifi electronic microscope, iPad (and appropriate apps), hand lenses, butterfly kits, access to the internet, for further research

Health and safety:

Wash hands after handling caterpillars.

Key information:

The Butterfly Conservation website (http://butterfly-conservation.org) is an excellent resource and includes a webcam showing live footage of a very busy butterfly habitat.

It also includes details of the annual Big Butterfly Count (http://www.bigbutterflycount.org) and schools are encouraged to submit their own data and make a contribution to the national database. Localised data sets are available for 2012 and 2013, so that children can see an emerging picture of butterfly sightings in their immediate area.

National curriculum links:

Describe the differences in the life cycles of a mammal, an amphibian, an insect and a bird; recognise that living things produce offspring of the same kind, but normally offspring vary and are not identical to their parents

Working scientifically links:

Recording data and results of increasing complexity using scientific diagrams and labels, classification keys, tables, scatter graphs, and bar and line graphs

Learning intention:

To observe life cycle changes closely in a controlled environment, making comparisons and suggesting reasons for similarities and differences between findings from controlled environment investigations and those recorded from observations in the wild

Scientific enquiry type:

Grouping and classifying

Success criteria:

- I can make detailed observations of the life cycle of a butterfly, identifying similarities and differences between the stages.
- I can identify physical and developmental changes that take place as a result of metamorphosis.
- I can record the number of individuals that fail to develop from egg to adult.
- I can describe the potential impact of various threats, including adverse weather conditions, predators and removal of habitat, on the butterfly population.

EXPLORE:

Prompt children to think back to their learning in Year 5 about the life cycles of insects and of butterflies in particular. They may have previously produced posters or notes about the life cycle of a butterfly, which could be referred to.

Show the Butterfly metamorphosis time lapse video (Video 1).

Ask: *What life cycle stages does any butterfly go through? What happens at each stage? When does the biggest change take place? Are there any similarities between the caterpillar and the final butterfly? What food do they eat? What physical features do they have? How many legs?*

Ensure that children recognise the magnitude of the change that occurs within the chrysalis. The adult butterfly is almost completely different from its earlier self.

ENQUIRE:

Explain to children that they are going to spend time over the next 4 or 5 weeks observing in detail the changes that caterpillars go through as they metamorphose to become adult butterflies. The caterpillars will continue to grow for about 2 weeks before each forms a chrysalis. Ensure that children have time to return repeatedly to the caterpillars (at least twice a week), measuring them and making notes about the changes they observe. Challenges 1 and 2 provide a pathway for understanding through children's experiences as they set up the butterfly house and observe the stages of the butterfly life cycle. Challenge 3 requires children to broaden their understanding of butterflies and of how this fragile life cycle can be impacted on negatively by external variables, such as the weather and predators.

Challenge 1: Children observe butterflies as they go through their life cycles and record detailed observations in their Changing World diaries

Ask the children to set up the butterfly house in the classroom and to observe the early stages of caterpillar development. (Ensure that there are sufficient caterpillars for the children to observe them in detail, before they are placed in the butterfly house.)

Ask: *What is the first stage in the butterfly life cycle? Where would the eggs have originally been laid? What do these caterpillars look like? What species of butterfly are they?*

Ask the children to make detailed drawings of the appearance of their caterpillar in their Changing World diaries.

Ask: *How long are the caterpillars? What physical features do they have? What food would they naturally eat? What are they feeding on now? How long will it take for them to develop to full size? How big do you think they will be? What will be the next stage in their development?*

Explain to the children that they need to make several types of observations and recordings as the caterpillars go through their life cycle. These include: counting the number of caterpillars that are placed in the butterfly house; counting the number of caterpillars that turn into pupae; and counting the numbers of butterflies that emerge from pupae.

After the life cycles have completed and the adult butterflies have been released into the wild, ask the children to present their data and interpret their results, giving reasons for what they have found out.

Challenge 2: Children make detailed observations and recordings of butterfly life cycles, and present their data graphically and in their changing world diaries

Ask the children to make detailed observations of the later stages in butterfly development. They should take measurements and make notes in their Changing World diaries about the changes they observe. Ask the children to present their findings in a graph.

Ask: *How much did the caterpillars grow? How big were the caterpillars just before they changed into chrysalises? Did all the caterpillars change into a chrysalis at the same time? Do all the caterpillars turn into chrysalises? What do you notice about the appearance of the chrysalis? Is it bigger or smaller than the caterpillar was just before it changed?*

The chrysalis will remain unchanged for 2 to 3 weeks. Ask the children to record all the details they can about the hatching process and how many butterflies emerge.

Ask: *Do all the chrysalises turn into butterflies? Do all the butterflies emerge from the pupae at the same time?*

Advise the children that they should look carefully for signs of change in the butterfly house caterpillars and ask them to record in detail what they see as the adult butterflies emerge and unfurl their wings. The act of pumping fluid through their wings straightens out the folds and brightens the colours.

Challenge 3: Children investigate the impact of external variables on butterfly populations

Explain to the children that they are going to explore the impact of external variables on the butterfly population, by using existing data from previous years. Direct them to the Big Butterfly Count data for 2012 and 2013 (http://www.bigbutterflycount.org/) and the UK Met Office's weather data (Resource sheet 1).

Ask the children to look at the data that they have available for 2012 and use to the data to try to explain why the summer weather in that year was very bad for butterflies. Ask them to compare the Met Office weather data for July and August, and sightings of butterflies recorded on the Big Butterfly Count website for 2012.

Ask: *Where do you think the butterfly population would have been hit the hardest? Why might that have been the case? Was there an increase in sightings of all butterflies in 2013? Why might this be? What do you think would be the perfect climate conditions for butterflies? What is more damaging to butterflies – very wet weather or unseasonably cold weather?*

REFLECT AND REVIEW:

Ask children from each of the challenge groups to share what they have found out with the rest of the class. Encourage them to think about how their observations, counts and measurements of butterflies in the butterfly house and in the school grounds, and the records of butterfly populations can help them to explain why the numbers of butterflies may vary from year to year.

Ask: *Do all caterpillars become butterflies? Can you suggest why not? What conditions are best for butterflies? What impact did the weather conditions in 2012 have on butterfly numbers? What other things could threaten butterfly species? How can humans improve the chances of butterflies successfully completing their life cycles?*

Ensure that children recognise that cold weather, and particularly wet weather, are very harmful to butterflies. Cold weather is damaging and if flowers fail to bloom in abundance there is less food for the butterflies. Other variables, such as the removal of suitable habitats and food plants , can also impact on butterfly populations. For example, the 'large blue' butterfly became extinct in the UK and the 'holly blue' butterfly was struggling until recently, but the position for both is now much improved. Holly blues feed on ivy and holly flowers, and are seen relatively frequently. The large blue has been reintroduced into the UK successfully, using stock from Sweden.

Ask children to suggest ways in which we could help butterflies. The BBC guide found at http://downloads.bbc.co.uk/wildaboutyourgarden/images/get_wild_about_your_garden_pocketguide.pdf includes plenty of accessible ideas, including the types of plants that attract different types of butterflies.

Ask: *Have our findings about butterflies helped us to understand why some animals are able to survive and others are not?*

Encourage children to consider the work they have done in the Module 4, Everything Changes, on variations in organisms and how living things survive.

EVIDENCE OF LEARNING:

Listen to children's responses throughout the lesson. Are their observations of the life cycle of a butterfly sufficiently detailed, identifying similarities and differences between the stages of the life cycle? Can they identify and describe the physical and developmental changes that take place as a result of metamorphosis? Can they suggest reasons why not all caterpillars become butterflies? Are children able to describe the impact of various threats, including adverse weather conditions, predators and the removal of habitat, on the butterfly population? Can they make suggestions, based on their research, about ways in which we can support and encourage a healthy butterfly population in the wild? Can they relate their observations to their work on variation in living things and the survival of organisms?

CROSS-CURRICULAR LINKS:

This lesson can be linked to Mathematics through the opportunities for collection, analysis and presentation of data in different ways.

OUR CHANGING WORLD

Key vocabulary:

nest, brood, fledgling, chick, young bird, juvenile, plumage, habitat, diet, migrate, migration, resident

Resources:

Binoculars, monoculars, video cameras, digital cameras, iPads with appropriate bird ID app, webcams, access to the internet for children to view bird information websites

Key information:

Animals are suited to their environments partly because of their physical adaptations, but also in part because of their behaviour patterns, for example, birds have beaks that are adapted for collecting and eating particular food types, but they also often feed at particular times and in different ways, for example, on the ground or hanging from tree branches.

LESSON 4: HOW DOES THE NUMBER, TYPE AND BEHAVIOUR OF BIRDS FOUND AROUND OUR SCHOOL CHANGE DURING THE YEAR?

LESSON SUMMARY:

During these lesson children observe the variety of bird life located around their school and within the local area throughout the year and identify patterns in the data they collect. By the end of these lesson children should be able to identify a variety of common birds and describe their life cycles, diets, adaptations and behaviour patterns. They interpret the data they collect, identifying and suggesting reasons for the patterns they see. They may, for example, recognise that some birds are native to the UK (they live here all through the year) while others may be migrants, arriving in and leaving the UK at certain times of the year.

These lessons link to Circle of Life and Reproduction modules in Year 5, and the Everything Changes module in Year 6.

National curriculum links:

Describe the differences in the life cycles of a mammal, an amphibian, an insect and a bird; identify how animals and plants are adapted to suit their environment in different ways and that adaptation may lead to evolution

Working scientifically links:

Reporting and presenting findings from enquiries, including conclusions, causal relationships and explanations of and degree of trust in results, in oral and written forms such as displays and other presentations

Learning intention:

To identify and make detailed observations of bird life in the local environment, looking for patterns in evidence collected over time

Scientific enquiry type:

Pattern seeking

Success criteria:

- I can identify and make detailed observations of birds found around my school or in the local area.
- I can use my knowledge of the life cycles of birds to help explain my observations of birds.
- I can identify patterns in the data I have collected, giving reasons why different numbers of birds might be observed at different times of year in the school grounds and in other locations further afield.
- I can investigate behaviour patterns of birds, for example their feeding habits, suggesting reasons for the behaviour.
- I can describe ways in which birds are adapted (both physically and behaviourally) to suit their environment.

EXPLORE:

Explain to children that the Royal Society for the Protection of Birds (RSPB) carries out an annual survey of the bird population to check on bird health and wellbeing. A very significant amount of the data collected comes from schools. Explain to children that this year they are going to monitor bird life in the school grounds, set up feeding stations and nest boxes, and provide data for next year's Big School's Bird watch database.

Show children the RSPB Top 15 birds list (Interactive 1) on the whiteboard. Go through the birds in turn and ask children to sort the list, identifying initially those birds that they have seen and those that they have never seen. Use the printed off Bird ID cards (Resource sheet 1) to display children's sort. This provided a useful focus for children's observations over time, allowing them to display the data that they collect and other observation outcomes.

Remind children that the images of the birds in the RSPB list are not to scale and that magpies, carrion crows and wood pigeons are relatively large birds; blackbirds and starlings are medium sized birds; and the chaffinch and robin are small birds. Children might benefit from creating paintings or drawings to scale of birds seen frequently around school.

Ask: *Which of these birds do you think that we will see most often when we start to make our observations? Are there any birds that are not on the RSPB list that you think you have seen in the playground or on the school field? Can you describe them? How could we find out what they are called?*

ENQUIRE:

Explain to children that during this lesson they are going to find out much more about the birds that visit the area around school. They are going to be logging observations regularly, collecting data and interpreting data as they go.

Discuss with children how, as a class, they might record and present their data, explanations and conclusions. These could be added to an Our Changing World display wall that is built up over the year and which could be supplemented by newsletters, blogs and tweets (if allowed by the school policy on use of social media).

Depending on the setting of the school, the school playground and school field might provide sufficient scope for making observations. As an alternative, children might make a planned visit to a local park or field study centre. Using binoculars, if available, allows children to make observations in more detail.

The challenges provide a variety of learning experiences for children. In Challenge 1 children identify and log the variety of birds observed in the school grounds. Challenge 2 requires children to identify patterns in sets of data, and Challenge 3 asks children to investigate the feeding habits and requirements of different types of birds.

Challenge 1: Children make and record observations of birds

Ask the children to to use the RSPB list to help them to recognise and identify birds in the school grounds. Provide the children with a Bird tally chart (Resource sheet 2) on which to note the birds they see as they follow a route around the school grounds. After they have made their initial observations, give them time to review what they have seen.

Ask: *What birds did you see? How many? Did you spot any unusual birds? Did you see any birds that are not on the Top 15 list? Did you see any parent birds collecting food to feed their young?* (This will be dependent on the time of the year.)

Encourage the children to develop a more rigorous approach to monitoring the birds that visit the locality during the year. For example, in addition to a walk around the grounds, they could decide on an observation station and then plan a viewing schedule (say, 30 minutes twice a day for 2–3 days, four times in the year).

Ask: *What do you want to find out? What data do you need to collect? Who will carry out the observations? How will you record your data? How will you analyse your data?*

Encourage the children to review the data that they have collected from multiple walks around the school grounds or from their observation station records.

Ask: *Can you identify any patterns in your data? What are your conclusions and can you justify them?*

Challenge 2: Children interpret patterns in bird observation data

Explain to the children that a group of pupils from Merryhills Primary School collected data about birds over a whole school year (Merryhills' bird observation data; Resource sheet 3). Observations were made on the second Thursday of each month and the children watched birds at their feeding stations for an hour, recording which type of bird was most frequently seen during that hour.

Ask the children to compare the school's data with information from the Big Schools' Birdwatch (Resource sheet 4). Big School's Birdwatch took place during the last two weeks of January, so children need to compare the Merryhills school data for January. Ask them to draw conclusions, based on the patterns that they can see across both sets of data.

Ask: *Which bird was seen most frequently by the Merryhills schoolchildren, and which was the most common in the Big Schools' Birdwatch? Is it the same bird? How does the variety of birds compare between the two sets of data? Blackbirds were the most frequently seen bird nationally, so why do you think that the number of blackbirds recorded in this school's data varies across the year? What do you notice about the pattern in the data for swallows? Why might this be? How might the weather affect the numbers of different types of birds observed? Would certain types of food attract particular birds and not others?*

Explain to the children that they are going to repeat Merryhills' investigation (this could be linked with Challenge 1) and that they are going to compare their findings with those found by the children at Merryhills. (Keep the data that the children collect because in future years a comparison can be made with your school's data, which can also be compared with the RSPB survey data. Some of the children might look up and analyse the RSPB School's Birdwatch data for the last 5 years and look for trends in the bird populations.)

Key information:
The following link provides useful information on feeding garden birds: http://www.rspb.org.uk/advice/helpingbirds/feeding/whatfood/index.aspx

Challenge 3: Children investigate feeding behaviours in birds

Explain to the children that they are going to investigate the feeding behaviours of birds. Provide access to the internet so that they can look at the RSPB website and/or other sources of information to find answers to some questions.

Ask: *What different types of food do birds eat? What foods do the RSPB Top 15 birds enjoy? What food do we provide at our feeding stations and bird tables? Do all birds feed on the same types of food? Do some birds feed on different foods at different times of the year?*

They should use what they find out to plan a 'feeding test' and write their plan in their Changing World diaries. The children can carry out the test at different times of the year, to establish which birds are attracted most frequently to what types of food and where the bird feeds, for example, is it on the ground? Stress to the children that they need to make their investigation rigorous and so need to plan carefully. Ask them some questions to prompt their thinking.

Ask: *What do you want to find out? What data do you need to collect? Who will carry out the observations? How long will your observations be? How often will you make your observations? What types of food will you put out? How will you record your data? How will you analyse your data? Can you identify any patterns? What are your conclusions and can you justify them?*

When the children have made sufficient observations, encourage them to think about what their data tells them about the feeding habits of different birds.

Ask: *Did some birds feed on the ground? Did some birds 'swoop in' and take food to eat elsewhere? Why do you think different types of birds feed in different ways (for example, display different patterns of feeding behaviour)?*

Suggest to the children that they might like to follow an experimental line similar to that of Charles Darwin, noting carefully which foods different types of birds feed on and then looking carefully at the shape of their beaks.

Ask: *Is there a pattern? Do birds with the same shaped beaks feed on the same types of food? Why might this be?*

Key Information:
Feeding behaviour is important because it can help animals to reduce the level of competition with other animals (for example, feeding on different foods means less competition for food) and to reduce the risk of being eaten by predators. This increases the chances of survival.

REFLECT AND REVIEW:

Encourage children to feed back what they have learned during the challenges. Ask those who have made observations to summarise their findings.

Ask: *Which of the RSPB Top 15 birds did you see? Did you observe any birds that were not on the list? Were you able to identify them? Were there any birds that you were not expecting to see? Any migrant birds that were 'just visiting'? Might you see different birds at different times of the year? Why? Did you find any evidence of how the birds are suited to their environment? How do you think they were adapted to make them suited to the environment? Did you see any patterns in their behaviour that help them to survive?*

EVIDENCE OF LEARNING:

Listen to children's responses as they make observations and interpret data throughout this lesson. Can they identify and name birds, making sufficiently detailed observations of birds frequently seen around the school grounds? Can they use their knowledge of life cycles in birds to help explain their direct observations and the patterns that they notice in data? Can they explain why different numbers of birds might be observed at different times of the year in the school grounds? Can children look at data from other locations further afield (such as information from the RSPB database) and suggest why the numbers of birds seen varies in different parts of the country? Can they identify and explain ways in which the birds are suited to their environment and suggest physical and behavioural adaptations?

OUR CHANGING WORLD

© LESSON 5: WHAT HAPPENS TO INVERTEBRATES DURING THE YEAR?

Key vocabulary:

invertebrate, justify, analyse, adaptation, predator, prey, survival, habitat, mollusc, worm, snail, woodlouse, centipede, millipede, shield bug, beetle, fly, aphid, names of other invertebrates identified during the lesson

Resources:

Materials for catching invertebrates and setting traps, digital cameras, video cameras, iPads with invertebrate identification apps, nets, sieves, large white sheets, binocular microscopes, magnifying lenses, secondary sources of information, for further research

Health and safety:

Wash hands thoroughly after handling animals or digging in soil.

Key information:

Invertebrates form a very diverse group of animals and occupy a wide range of habitats. They show many adaptations to make them suited to their environments. These adaptations include body structure, colour, mouthparts for feeding, life cycles and behaviour patterns.

LESSON SUMMARY

During this lesson children build on their previous work in Years 3 and 4 on invertebrates (minibeasts) and their habitats. Children monitor invertebrates found in the locality of the school during the year, identifying and recording the numbers of different types of invertebrate and their habitats. They conduct some simple investigations into the behaviour of invertebrates and examine ways in which these animals show adaptation to the environments in which they live.

By the end of these lesson children have become familiar with a range of invertebrates, and they are able to identify them and suggest ways in which they are suited to their environment.

Preparation required:

Materials and equipment is needed for making some invertebrate traps. (Some ideas can be found in Resource sheet 1.) Also, collect some live invertebrates, in individual containers, prior to the start of the lesson. Children completing Challenge 2 need to be able to observe the behaviour of earthworms and woodlice closely. You may wish to set up or purchase a wormery and/or woodlouse choice chamber (Resource sheets 2 and 3).

National curriculum links:

Describe how living things are classified into broad groups according to common observable characteristics, similarities and differences, including micro-organisms, plants and animals; identify how animals and plants are adapted to suit their environment in different ways and that adaptation may lead to evolution

Working scientifically links:

Reporting and presenting findings from enquiries, including conclusions, causal relationships and explanations of and degree of trust in results, in oral and written forms such as displays and other presentations

Learning objective:

To extend understanding of different types of invertebrates, with particular reference to the ways in which they are suited to the environments in which they live

Scientific enquiry type:

Observing changes over different periods of time

Success criteria:

- I can identify and make detailed observations of a variety of invertebrates found around my school.
- I can describe ways in which different invertebrates are adapted to their environments.
- I can suggest reasons why the adaptations of invertebrates help them to survive.
- I can design, plan and carry out investigations with invertebrates.
- I can present my results effectively and give reasons to support my conclusions.

EXPLORE:

Provide children with some examples of invertebrates (in individual containers) that have been collected from around the school locality. You can supplement the live specimens with photographs of different types of invertebrates (Slideshow 1).

Ask children to compare and contrast the features of these animals and to identify them.

Ask: *In what kinds of places do you think these animals live? In what ways do you think they are suited to the places in which they live? What do you think happens to them at different times of the year?*

ENQUIRE:

Explain to children that they are going to explore what happens to some of these invertebrates during the year and investigate some of the ways in which they are adapted to their habitats.

The challenges look at different aspects of invertebrate life and life cycles and are differentiated according to the focus of the study and the skills needed to complete it. In Challenge 1 children monitor how the numbers and types of invertebrates found around the school grounds change during the year and they try to link different types of invertebrates to particular habitats. In Challenge 2 children look at two particular groups of invertebrates (for example, earthworms and woodlice), monitor their numbers and investigate some of their behaviour patterns. Challenge 3 requires that children investigate different types of trees as invertebrate habitats and examine differences in the invertebrate populations in these habitats, and also during the course of the year. In all the challenges children need to decide on the ways to present their findings and how to explain their conclusions.

For children undertaking Challenge 2 to be able to investigate the behaviours of earthworms and woodlice more closely, they need to conduct some specific experiments. Setting up a wormery is a very good way of observing the behaviour of earthworms (see Resource sheet 2, for instructions).

The behaviour of woodlice can be tested by setting up a 'choice chamber' and by controlling the conditions in different areas to build a picture of the type of habitat preferred by woodlice. It is possible to purchase pre-made choice chambers, but instructions on how to make one are provided in Resource sheet 3.

Explain to children that they should also work in pairs to become 'experts' on a particular type of invertebrate. Ask them to choose an invertebrate to study and explain that they should produce an invertebrate passport that records all its important features.

Challenge 1: Children investigate invertebrates and their habitats

Explain to the children that they are to going to become experts on the environments in which invertebrates live. Suggest that they should start by looking around the school grounds for places where they might find invertebrates and ask them to record what they see in those places.

Once the children have done this, ask them to select 3–5 different types of places (for example, in the soil, under stones, a rotting tree trunk, a bush and piles of dead leaves) and explain that they will return to these places at different times during the year (perhaps four visits). They need to decide how they are going to trap, collect, count and identify specimens of each type of invertebrate so that they can observe them more closely. Confirm that there will be traps and other suitable equipment (for example, a sieve for sorting soil or leaf litter, hand lenses and binocular microscopes) and explain to the children that at each visit they should record the weather and features of the habitat. This includes noting what the temperature was, whether it was wet or dry and where the animals were (in the dark or light, for example).

Remind the children as they make their visits that they should present their results in appropriate ways so that they can try to answer questions.

Ask: *What kinds of invertebrates did you find? How many of each type? Which habitats were they in? Which ones did you find in more than one place? Did the numbers change during the year? Were you able to identify ways in which the animals might be adapted to their habitats?*

Challenge 2: Children investigate invertebrate behaviour

Explain to the children that they are going to study two invertebrates in detail and, in particular, investigate some of their behaviour patterns. (Woodlice and earthworms are particularly good for this and are used as examples, but other invertebrates, for example snails, could be studied in a similar way using different experiments.)

Ask the children to carry out a survey of the school grounds and to identify the different places where they can find earthworms and woodlice. Explain to them that they need to decide how they are going to trap and count the number of each type of invertebrate in each place. At each visit they should record the weather, temperature, features of the habitat (for example, whether it was wet or dry) and note if the animals were in the light or the dark. Let the children know that they will visit 3–4 times during the year.

Ask the children to collect some earthworms for the wormery so that they can observe earthworm behaviour more closely, and to design and carry out some experiments on woodlouse behaviour using a choice chamber.

Ask: *In what kinds of places did you find earthworms/woodlice? What were the conditions like? What did you find out about the behaviour of earthworms/woodlice from your experiments and observations? Did the classroom experiments help you to understand what you saw outside? Can you suggest ways in which earthworms/woodlice are suited to the environments in which they live?*

Key information:

Many different types of invertebrates live on trees and different types of trees have different kinds and numbers of invertebrates associated with them. The oak tree has one of the largest variety of invertebrates associated with it.

Challenge 3: Children investigate invertebrates that inhabit trees

Explain to the children that they are going to investigate, in some detail, invertebrates that live on trees, and record and explore the changes in invertebrate types and numbers during the year.

Ask the children to select two different types of trees. (For practical reasons of accessibility, these should not be too high so that the children are able to reach and shake the branches using a metre rule.) Ask them to record the type of tree, its size and other characteristics, such as the type of bark (for example, its colour and whether it is smooth or rough).

Explain to the children that they will visit their two types of trees three or four times in the year and during each visit they should first directly examine the trunk, branches and leaves for evidence of invertebrates, recording both the types and numbers of invertebrates that they find. Once they have done this they should spread out a large (preferably white) sheet under the branches of the tree and then shake the branches using a metre rule. Warn them not to stand directly underneath as things will fall down onto the sheet. Ask the children to examine what has fallen onto the sheet, and to record and identify examples of invertebrates. They can collect examples of different invertebrates for identification and further examination. Remind the children that some invertebrates move very quickly!

Once the examination has been completed, the sheet should be shaken to remove the animals and the process then repeated under another part of the tree. Ask the children to carry out this process two or three times for each tree.

Ask the children to use secondary sources of information to find out which types of trees have the greatest variety of invertebrates living on them.

REFLECT AND REVIEW:

Review with children their initial ideas and their findings as their investigations progress. Ask different groups to summarise what they have found out to date, sharing any difficulties that they have had and explaining how they have overcome their difficulties and what they are planning to do next.

At the end of the year, explain that you are going to hold a scientific conference at which the children are invited to present their findings. This could take the form of conference sessions at which children present their 'papers', with time for discussion after each presentation. Alternatively, you could organise a poster session (possibly coinciding with a parents evening or during a lunchtime), at which each of the groups prepares a display poster (A3 size or slightly bigger) to explain their investigations, findings and conclusions to invited parents, other pupils and teachers.

EVIDENCE OF LEARNING:

Listen to children's responses during this lesson. Do they make sufficiently detailed observations of the invertebrates and their habitats? Can they identify similarities and differences between the invertebrates? Can they relate the different types of invertebrates to their habitats and lifestyles? Do children give examples of how the physical adaptations and behaviours of the different invertebrates make them suited to the environments in which they live? Can they explain and justify how they carried out their investigations, and provide evidence to support their conclusions?

THE NATURE LIBRARY

INTRODUCTION

This is a challenging module in which children will build on their knowledge of living things from previous years and deepen their understanding of why and how organisms are classified. They will explore the process of classification in some detail and how it differs from, but relates to, the identification of living things. The structure, function and purpose of classification systems will be explored with specific reference to living things. Children will become aware of the types and characteristics of organisms that belong in each of the five kingdoms of living things (animals, plants, fungi, bacteria and Protista) and the major sub-groups the kingdoms include. Although they will devise their own systems of classification, children will learn about how Linnaeus developed the system for classifying all living things using their observable characteristics. They will be introduced to the idea of how scientists use 'conventions' in order to ensure that everyone means the same thing when they refer to, for example, an organism by its scientific name. This module links to the modules in Our Changing World.

When working scientifically, children will use observations and secondary source material to help classify living things, record plants and animals in the school environment and use evidence to support or refute ideas. They will use a range of approaches to present and communicate their findings to others including questioning themselves and their peers, evaluating the strength of evidence used to support arguments.

Key vocabulary:

General terms: identify, identification, classify, classification, division, family, genus, species, reason, common characteristics, distinguishing characteristics, leaves, shape, size, colour, backbone, wings, jointed legs, cased, transparent, antennae, shell, segments, explain, group, small, harmful, beneficial (helpful), colony, colonies, mould, multiply, historically, grouping, Aristotle, Carl Linnaeus, kingdom, Phillip Miller, John Ray, botany, conventions

Kingdoms of living things: Animalia, Plantae, Fungi, Protista, and Monera

Plant kingdom: flowering plants, conifers, ferns, mosses and algae

Animal kingdom: vertebrates, fish, amphibians, mammals, birds, reptiles, invertebrates, molluscs, annelids, arachnids, insects, arthropods

Micro-organisms: (3 kingdoms: Fungi, Monera, Protista), micro-organisms (microbes) bacteria

Extension vocabulary:

Vocabulary for working scientifically: question, investigation, fair test, change, measure, predict, prediction, explanation, observations, draw conclusions, plan, do, review, risk

FACT FILE:

Classification is not the same as identification. During classification the emphasis is on the similarities of objects in order to demonstrate that they belong to the same group. Identification focusses on the differences between objects in order to be able to give a specific name to that particular thing. The two processes are linked but not interchangeable.

Classification depends on developing groups and subgroups of objects at different levels. All the objects to be classified form the first set, for example, living things. These divided in sub-groups, for example the 5 kingdoms, each of which is subdivided into the next set of subgroups and so on until it is not possible to sub-divided things any further. This can be represented in two ways as shown in Figures 1 and 2. Simplified classification charts for plants and animals are included with the assets for the module.

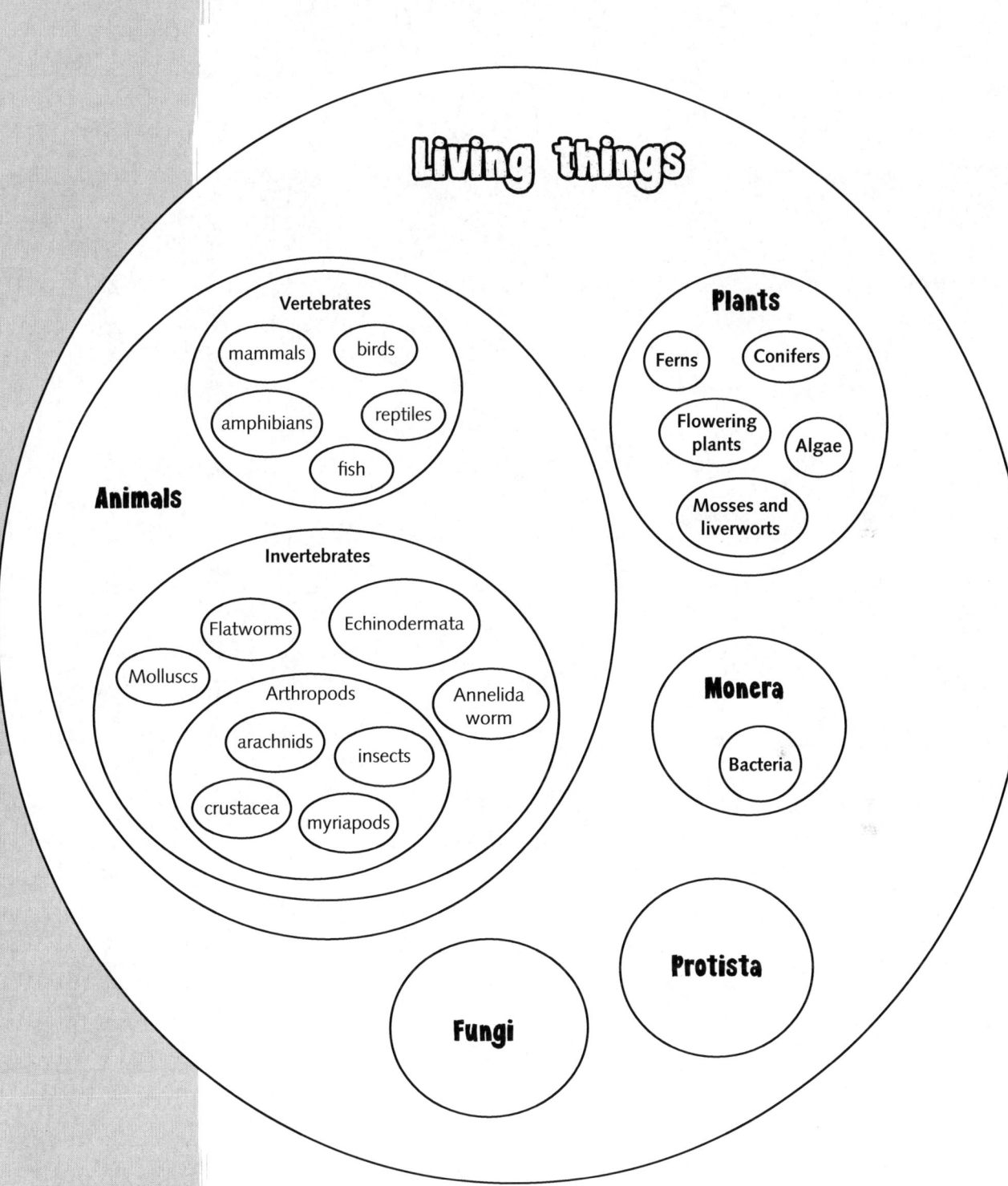

Figure 1 Classification of living things shown by a set-diagram

Classification systems vary depending on their purpose and can be changed when new evidence comes to light. The main system for classifying living things was originally developed by Carl Linnaeus in the 19th Century and, although it has been modified, it is still used today.

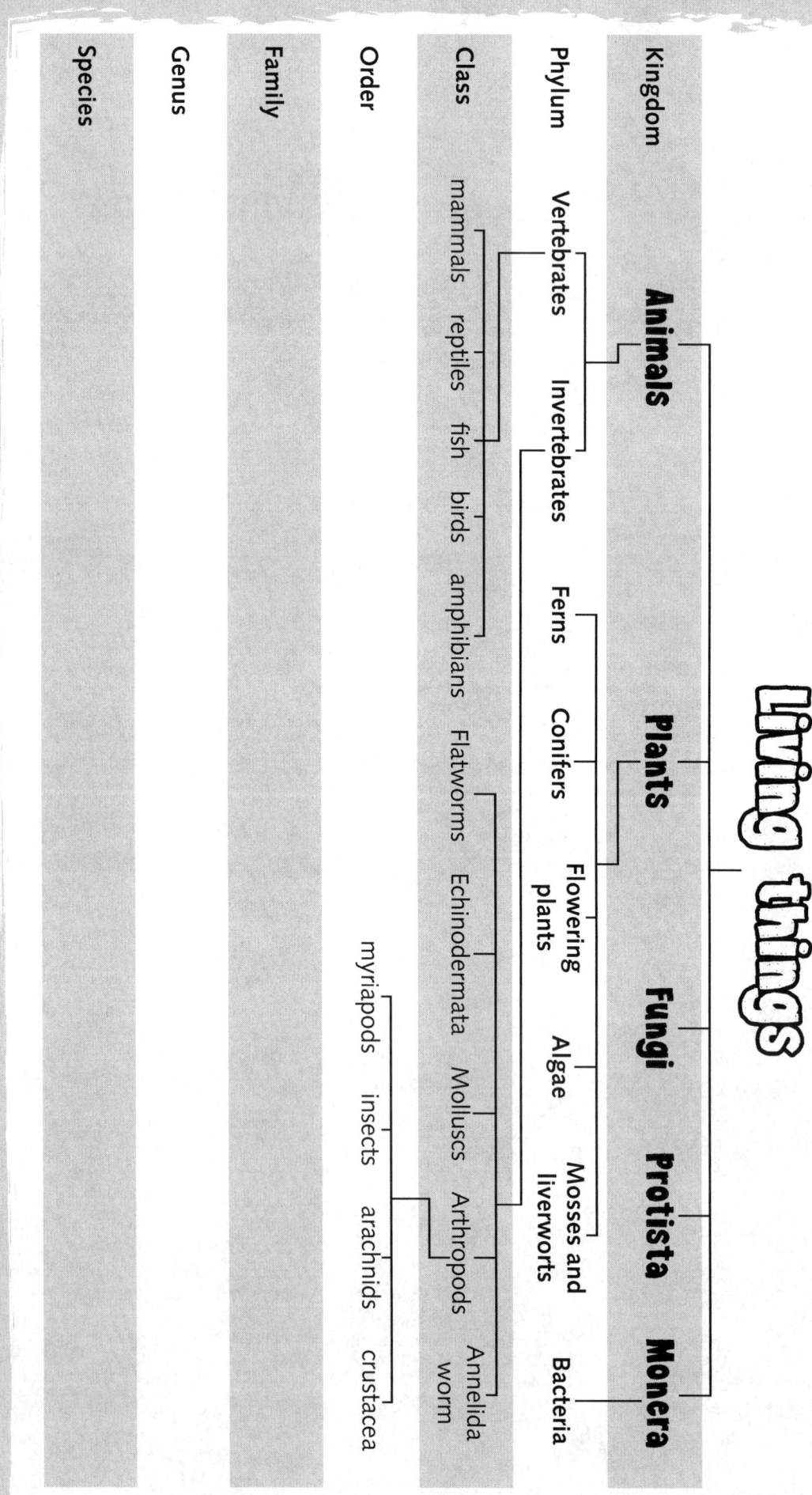

Figure 2 Structure of a classification chart for living things

Plants are classified into 5 main groups (Divisions) and then subdivided at different levels. The main levels are; Kingdom, Division, Order, Family, Genus, and Species. Animals are classified into 2 main groups (vertebrates and invertebrates) and then at different levels in a similar way to plants. The main levels for animals use a slightly different terminology and are: Kingdom, Phylum, Class, Order, Family, Genus, and Species. Every type of plant or animal can be described in terms of these levels. Scientists classify things because they want a system which includes all organisms and, importantly, all scientists are able to use the same groups consistently across the world. Table 1 gives an example of how a particular plant and an animal might fit into these levels.

Table 1 Classification for a daisy and a leopard

A daisy, scientific name: *Bellis perennis*		A leopard, scientific name: *Panthera pardus*	
Kingdom	Plantae (plants)	**Kingdom**	Animalia (animals)
Division	Angiosperms (flowering plants)	**Phylum**	Chordata (vertebrates)
Order	Asterales	**Class**	Mammalia (mammals)
Family	Asteraceae /Compositae	**Order**	Carnivora (carnivores)
Genus	*Bellis*	**Family**	Felidae (cats)
Species	*perennis*	**Genus**	*Panthera* (big cats)
		Species	*pardus*

The convention for naming organisms is that it should have two parts made up of the genus to which it belongs (a bit like a surname) and the species name (a bit like a given name). The genus is determined by matching the organism to the appropriate group as determined by the classification system being used. The species name is given specifically to the organism in question so often is named after the person who found it or one of its features. Together the genus and species provide a unique name for the organism.

The terminology for these living things is complex and not always consistent in reference materials, especially older sources. In this module we will refer to micro-organisms or microbes to include the three remaining kingdoms of living things which are: Fungi (moulds and mushrooms), Protista (single-celled organisms that have a nucleus e.g. amoeba) and Monera (singled-celled organisms which do not have a nucleus and include bacteria).

Linnaeus could only base his classification on observable features of organisms but scientists today are able to use improved optical, digital and electron microscopes as well as DNA analysis and sequencing, among other techniques, in order to determine which organisms are closely related. The results indicate that, although the overall system remains sound, some organisms may be in the 'wrong' groups. Of course this is not helpful if you are looking at a plant or animal in the middle of the jungle; so a classification system based on observable features is still very important.

There are many things in science that scientists may disagree about but they have ways of resolving them. The agreements on what is acceptable are usually called 'conventions' so that all scientists understand what has been agreed. Classification of living things has seen many changes since Linnaeus devised his system. Each change, big or small, has to be agreed internationally before it is implemented – this can take years as arguments are put for and against the change.

Common misconceptions:

Children often think that:

- the groups of living things are completely independent e.g. 'a fish is a fish, not animal' because they are not aware of how the groups relate to each other
- fungi are plants but they are not because fungi cannot make their own food by photosynthesis.

THE NATURE LIBRARY

C

LESSON 1: CAN YOU SORT THIS MESS?

LESSON SUMMARY:

In this lesson children build on their knowledge from previous years about how living things can be grouped together in different ways according to the characteristics they have in common; this is classification. By the end of this lesson children have begun to understand the process of classification. Children then apply the process of classification in the other lessons in this module.

National curriculum links:

Describe how living things are classified into broad groups according to common observable characteristics and based on similarities and differences, including micro-organisms, plants and animals

Learning intention:

To demonstrate understanding of the process of classification

Scientific enquiry type:

Grouping and classifying things

Working scientifically links:

Recording data and results of increasing complexity using scientific diagrams and labels, classification keys, tables, scatter graphs, and bar and line graphs

Success criteria:

- I can develop a classification system for sweets.
- I can explain how and why I have classified certain objects in certain ways.

EXPLORE:

Provide children, in groups of three or four, with Living things (Resource sheet 1). The cards include vertebrates, invertebrates, flowering plants, ferns, mosses, fungi and bacteria. Ask them to sort them in any way they like. Each group then explains to another group why they have sorted the images in a particular way. Repeat this activity, asking each group to sort their images using different criteria and to then explain why they have done it.

Ask: *What groups do we put living things into? Why do you think we need to sort living things into groups? Do you think it is easy?*

At this stage, the discussion is to get children thinking about why living things are grouped and the types of characteristics that can be used. It is a good time to introduce the term 'classification' to describe the placing of objects into groups based on their shared characteristics.

Ask: *Why might we want to classify something? When have you classified things in other learning experiences at school or in real life?*

ENQUIRE:

Explain to children that they are going to develop their own classification system, not for living things, but for different types of sweets. Children work in groups of two or three.

Explain to children that their challenge will be to create a classification system for sweets. Differentiation is provided by the support given and by the number of criteria and levels of classification children are asked to use. For example, a criterion might be 'made of chocolate' or 'contains toffee'. The number of levels process is very similar to creating a series of sets with sub-sets; for example, the first (top) level is sweets; the second level might contain two sub-sets: toffees and chocolates; at the third level, the toffees might be divided into three sets: toffee only; toffee wrapped in chocolate; and toffee wrapped around chocolate. Children may be familiar with the process of setting from Maths, so it will be helpful to make the link here.

Key vocabulary:

identify, identification, classify, classification, reason, common characteristics, distinguishing characteristics

Resources:

A large selection of different types of sweets (toffees, chocolates, marshmallows, peppermint creams, liquorice allsorts – some of the selections produced by various manufacturers would be suitable), marker pens, flip chart paper, hoops of different sizes (alternatively, different lengths of string that can be tied to form loops of different sizes)

Key information:

Children should know that living things belong to different groups (such as animals: vertebrates, invertebrates; plants: flowering plants, ferns, mosses) but they will not be aware of all of the groups nor of how they relate to each other. This lesson and the following lessons demonstrate how the classification of living things has come about.

Key information:

Make children aware that classification is not the same as identification. During classification the emphasis is on the similarities of objects which demonstrate that they belong to the same group. Identification focuses on the differences between objects in order to be able to give a specific name to that particular thing. The two processes are linked but not interchangeable.

Key information:

There are no restrictions on the number of sets and sub-sets, but for practical reasons children should be encouraged not to create sets in which there are fewer than five sweets to begin with. Obviously, as sub-sets of sub-sets are formed, the numbers of sweets in each sub-set decreases.

Health and safety:

When using sweets and other foodstuffs in the classroom, remind children not to eat anything unless told they may do so and that they should wash their hands after handling sticky things. See Be Safe! section 7.

Challenge 1: Children group 20–30 sweets, record their common characteristics and divide one group into sub-sets, recording common characteristics

Provide each group of children with a pile of between 20 and 30 mixed sweets. Ask them to think of ways they could group the sweets by putting those with the same characteristics together. Ask them to record their ideas by putting each group of sweets into hoops or circles drawn on a flip chart. They should list the characteristics the sweets in each group have in common.

As the children become confident in putting sweets into sets, challenge them to look at the sweets in one of their sets and to put them into sub-sets. They then record the features that the sweets in each sub-set have in common. Use the flip chart to show the sets and sub-sets the sweets have been put into.

Ask: *What features do the sweets in each set/sub-set have in common? Why did you choose those features? What name would you give to each of the sets?*

Challenge 2: Children group 30–50 sweets, record their common characteristics and divide groups into two or three sub-sets on a poster

Provide each group of children with a pile of between 30 and 50 mixed sweets. Ask them to think of ways they could group the sweets by putting those with the same characteristics together. Ask them to record their ideas by putting each group of sweets into hoops or circles drawn on a flip chart. They should list the characteristics the sweets in each group have in common.

Challenge the children to develop a series of sub-sets at two or three different levels. They should record their sets and sub-sets on a poster for display and give each set, sub-set and level a name.

Ask: *What features do the sweets in each set/sub-set have in common? Why did you choose those features? What name would you give to each level? What do you notice about the types of sweets in the sub-sets at the lower levels? Is there any subset that has only one sweet in it? Do you think that there might be other sweets in the world that might belong in that subset? How would you find out?*

Challenge 3: Children group 40–60 sweets, record their common characteristics, divide groups into sub-sets and record on a classification tree

Provide each group of children with a pile of between 40 and 60 mixed sweets. Ask them to think of ways they could group the sweets by putting those with the same characteristics together. Ask them to record their ideas, putting each group of sweets into hoops or circles drawn on a flip chart. They should list the characteristics the sweets in each group have in common.

Challenge the children to develop a series of sub-sets at three or four different levels. They should record their sets and sub-sets in a 'classification tree' to show how and why they grouped the sweets at the different levels. Each level and set or sub-set should have a name.

REFLECT AND REVIEW:

Ask some children to explain their classification system for the sweets.

Ask: *Do you think children in other countries would classify sweets in the same way? Would it be helpful if they did? What might be the advantages and disadvantages?*

To help children think about the levels of classification, ask them to choose one sweet and set out its classification. For example:

	Level name	Set/sub-set name
Level 1	Food type	Confectionary
Level 2	Sweet type	Chocolates
Level 3	Centres	Soft jelly
Level 4	Wrap-around	Chocolate outside
Level 5	Product	Turkish delight

EVIDENCE OF LEARNING:

Listen and watch children as they sort the sweets and as they discuss the meaning of the words 'classification', 'levels', 'sets' and 'sub-sets'. Can children explain why they have grouped the items as they have? Are they able to recognise the common features and group sweets accordingly? Can they suggest why classifying things can be helpful? Can they explain some of the difficulties in devising a classification system? Are they able to suggest alternative ways of classifying? Can they present and explain their findings clearly?

THE NATURE LIBRARY

Key vocabulary:

identify, identification, classify, classification, leaves, plant, shape, size, colour, flowering plants, conifers, ferns, mosses, algae, kingdom, division, species

Resources:

Examples of different types of plants to include at least one moss, one fern, a conifer and a flowering plant, with photographs to increase the variety; sticky notes, mini whiteboards

Health and safety:

Refer to Be Safe! section 9 when handling plants. Be aware of any sap or pollen allergies in your class group.

Key information:

Classification systems are developed for a purpose. In the context of this lesson, garden centres want to make it easier for people to buy the right types of plants for their own gardens. Hence the grouping of plants depends on things like type of plant, where it goes, its shape and size. Scientists classify things because they want a system which includes all organisms and, importantly, that allows all scientists to use the same groups consistently around the world.

[C] LESSON 2: CAN YOU FACE THE GARDEN CENTRE CHALLENGE?

LESSON SUMMARY:

In this lesson children decide ways in which to group plants. They apply their classification skills to different types of plants, giving their reasons for the groups and justifying them to others. By the end of the lesson they recognise that plants can be classified in different ways and that scientists have a classification system that includes all plants.

Preparation required:

Ideally this lesson could be linked with a visit to a garden centre or someone from a garden centre could come into school to talk about what they do. This would need to be arranged in advance.

National curriculum links:

Describe how living things are classified into broad groups according to common observable characteristics and based on similarities and differences, including micro-organisms, plants and animals; give reasons for classifying plants and animals based on specific characteristics

Learning intention:

To apply the process of classification to plants

Scientific enquiry type:

Grouping and classifying things

Working scientifically links:

Recording data and results of increasing complexity using scientific diagrams and labels, classification keys, tables, scatter graphs, and bar and line graphs

Success criteria:

- I can group plants in different ways.
- I can explain why I have classified plants in different ways using the features they have in common.
- I can describe the classification system that scientists use to classify plants and name the main plant groups.
- I can use scientific vocabulary to classify plants.

EXPLORE:

Display the different types of plants so that all children can look at them in detail. Ask them to work independently to record different ways they could sort the plants. Ask what characteristics the plants might have in common that they could use to put them in the same groups. How might they use the characteristics to sort a large number of plants?

They could write each idea on a separate sticky note or as a list on a mini whiteboard. Encourage children to use their own knowledge about plants as well as what they can see in the examples.

Ask children, in pairs, to create a joint list, discarding repeats so they have each criterion once.

As a class, discuss all the different ways you could classify plants. Suggestions might include leaf shape, size, colour of leaves or flowers, growing conditions, the growth form (does it grow tall or spread out?) and whether it produces fruit.

ENQUIRE:

Display How do garden centres organise their plants? (Slideshow 1) and ask children to consider the question. Talk through the images on the subsequent slides and the ways in which the plants are organised. Encourage children to reflect on what they have already learned about classifying and grouping similar objects and species. Remind children that scientists classify for a purpose. Display slide 5 and explain to the class that their challenge is to help the staff at Sunny View Garden Centre to organise their plants in the best way possible ready for the visit of Professor G Reen. The challenges are differentiated by the ways in which children are asked to organise their plants and the level of detail they are asked to include in the labelling.

Note: It is recommended that the Garden centre floor plan resource sheets are printed out in colour from Collins Connect.

Challenge 1: Children work in pairs, with adult support where appropriate, to group approximately 20 plants according to the habitat in which the plant would grow best

Give each pair cut out images from Garden centre floor plan (Resource sheet 1). The children use the images provided to create a floor plan of the garden centre indicating which plants are placed where. This should be based on the habitat in which the plant would grow best. The Professor is going follow this route on his visit. Each child should create a label for two or three of the plants which states the common name and which scientific classification group it belongs to, such as mosses, ferns, conifers or flowering plants.

Challenge 2: Children work in groups of two or three to group approximately 30 plants scientifically

Give each group cut out images from Garden centre floor plan (Resource sheet 2). They should arrange the plants scientifically based on the discussion they have had in the Explore part of the lesson, such as leaf shape or height. They should agree the criteria for sorting the plants and consider where they are placed in relation to other groups of plants. They present a poster-sized floor plan that clearly indicates the names they have assigned to each area of plant types. The name of the group should indicate the reason for it being chosen and include scientific vocabulary. Suggested responses might include deciduous small green plants or pink flowering shrubs, whereas pink flowers would not be an acceptable reason. Each child should create a label for two or three of the plants that states the common name, the scientific name and the scientific classification group to which it belongs, such as mosses, ferns, conifers or flowering plants.

Challenge 3: Children work independently to group and classify about 40 plants according to criteria they decide

Give the children cut out images from Garden centre floor plan (Resource sheet 3). Ask them to arrange the images into a floor plan on poster paper that Professor G Reen can use as a map of the site during his visit. Ask the children to write a commentary for another member of staff at the Sunny View Garden Centre to use when they are showing the visitor around. The commentary should include the reason for placing the groups where they are, as well as the names of the plant groups. Suggested answers for why the plants were classified in a certain way will be similar to that in Challenge 2. Each child should create a label for two or three of the plants which states the common name, the scientific name and the scientific classification group it belongs to, such as mosses, ferns, conifers or flowering plants.

REFLECT AND REVIEW:

Ask children to share how they organised their garden centre and to explain to other groups why they were organised in certain ways.

Let children know that Professor G Reen was delighted with his visit and that he thought it might help if he sent a simplified classification chart for them to look at. Show Plant classification chart (Resource sheet 4). The chart shows how scientists classify plants to help them with their work. Discuss the chart with children, noting the levels of classification and the different groups with the examples of the types of plants that belong in each group.

EVIDENCE OF LEARNING:

Listen to children as they classify their plants and ask them to justify their reasons. Take note of their use of language, linked to plants and classification. Can children decide on ways to group their plants? Can children justify their grouping? Do they use scientific vocabulary in their justifications, such as "I have grouped these green plants together because their leaves are the same shape and they are all coniferous. They are next to the green plants that are deciduous." Are they able to produce accurate labels for the individual plants? Do they recognise that the classification system has different levels? Can they describe the five main groups in the classification of plants?

Key information:

Plants are classified into five main groups (Divisions) and then at different levels in a similar way to animals (see Lessons 3 and 4). The levels are: Kingdom, Division, Order, Family, Genus and Species. Every individual plant can be described in terms of each of these levels. In practical terms, for example in a garden centre, the most frequently used are Kingdom (plants), Division (algae, mosses and liverworts, ferns and horsetails, conifers, flowering plants) and then Family, Genus and species.

THE NATURE LIBRARY

Key vocabulary:

identify, classify, vertebrates, invertebrates, backbone, fish, amphibians, mammals, birds, reptiles

Resources:

Internet access, secondary reference resources

LESSON 3: HOW ARE VERTEBRATES GROUPED TOGETHER?

LESSON SUMMARY:

In this lesson and Lesson 4 children consider the classification of animals. After revising their knowledge of different types of animals from previous years, they investigate in more detail the grouping and classification of vertebrates. (Invertebrates are covered in Lesson 4.) By the end of this lesson children have a clear understanding of the differences and similarities between vertebrates and invertebrates, and the main groups of vertebrates.

National curriculum links:

Describe how living things are classified into broad groups according to common observable characteristics and based on similarities and differences, including animals; to give reasons for classifying animals based on specific characteristics

Learning intention:

To explore the classification of animals and recognise the main groups of vertebrates

Scientific enquiry type:

Grouping and classifying things

Working scientifically links:

Reporting and presenting findings from enquiries including conclusions, causal relationships and explanations of and degree of trust in results, in oral and written forms such as displays and other presentations

Success criteria:

- I can use appropriate vocabulary to classify animals and group vertebrates.
- I can use observable characteristics to group and classify vertebrates.
- I can apply what I know in order to classify an unknown animal.

EXPLORE:

Give each pair of children a cut up set of cards from Animal card sort (Resource sheet 1) to sort. Ask children to group the animals and be prepared to explain the criteria they have used. Once they have made their groups they should provide a name for each group which indicates why the animals are in that group.

Encourage children to think about groups based on combinations of characteristics; for example, if the grouping was simply on 'wings or no wings', they would end up with bats and birds in the same group. However, if the grouping was based on 'wings and feathers', the group would only contain birds.

After a few minutes, create a 'marketplace' where pairs visit other pairs to share and listen to their reasons for grouping the animals in a certain way. At this stage there is no need for children to use the 'scientific' classification of animals, but they should be encouraged to use their existing knowledge as appropriate. Emphasise the need to identify the characteristics that are the same for the animals to be included in a particular group.

Gather together as a class and record the most commonly used group names in the class during this activity. Show slides 1 and 2 from the Vertebrates and invertebrates slideshow (Slideshow 1). They should be familiar and confident with these words now.

Explain to children that their challenges are to look more carefully at the different types of vertebrates (fish, amphibians, reptiles, birds and mammals). The challenges are differentiated by the level of detail about the different types of vertebrates children are asked to notice and explain. All challenges have the same starting point.

Key information:

Animals are classified into two main groups (vertebrates and invertebrates) and then at different levels in a similar way to plants (see Lesson 2). The levels are: Kingdom, Phylum, Class, Order, Family, Genus and Species. Every individual animal can be described in terms of each of these levels. The everyday terms used to describe animal groups do not fall neatly into the scientific levels, so it is necessary to check.

ENQUIRE:

Show the children slides 3, 4, 5, 6 and 7 from the Vertebrates and invertebrates slideshow (Slideshow 1) and ask them to make notes of the common characteristics of that particular vertebrate using Who has what? (Resource sheet 2). When they are finished, allow them a few moments to underline the characteristics they think are unique to each type of vertebrate.

Challenge 1: Children create a set diagram using a template

Using the information they have noted from the slideshow and other secondary sources, ask the children to create a set diagram that shows which characteristics are common to all vertebrates and which characteristics are common to each of the sub-groups they have identified. The Set diagram template (Resource sheet 3) provides a template for this.

Challenge 2: Children create a classification chart

Using the information they have noted from the slideshow and other secondary sources, ask the children to create a classification chart that shows how vertebrates are part of the animal kingdom and that includes which characteristics are common to all vertebrates and which characteristics are common to each of the sub-groups they have identified. They should make sure that the different groups are at the correct levels of the chart but at this stage the children do not need to include invertebrates, which are covered in the next lesson.

Challenge 3: Children create a set of Top that cards

Using the information they have noted from the slideshow, secondary sources and the cards from Animal card sort (Resource sheet 1), the children work together to create a class set of cards for a variety of vertebrates. A template is provided in Top that party game (Resource sheet 4), but the children may prefer to make their own. Each set needs to include different types of fish, amphibians, reptiles, birds and mammals.

Each card must include a photo, the name of the animal, the classification group it belongs to, two characteristics that are shared by that type of vertebrate and four comparable facts that compare common observable and number-based statistics, such as average length, average number of young produced in a lifetime and average age or number of legs.

Children devise the rules for a game they can play together that helps them in classifying animals; for example, by swapping cards, the winner is the first to collect all the vertebrates that are fish.

REFLECT AND REVIEW:

Describe the following invented animal to children and ask them to do a quick sketch as you describe it. Ask them to then classify the animal into the group in which they think it fits best.

This animal has a tiny soft rounded body, it has a backbone, it breathes with lungs, lays eggs and has pink, purple and yellow feathers.

If time allows, repeat this activity. Children could be asked to think of their own descriptions.

EVIDENCE OF LEARNING:

Observe children as they are classifying their animals. Are they using the appropriate vocabulary? Are they asking and responding to each other's questions in a scientific way? Are they applying what they already know? Do they use observable characteristics to classify animals? Do they know the characteristics of the different groups of vertebrates? Are they more confident in classifying vertebrates?

THE NATURE LIBRARY

Key vocabulary:

invertebrates, wings, jointed legs, cased, transparent, antennae, shell, segments, classify, identify, molluscs, annelids, arachnids, insects, arthropods

Resources:

Internet access, secondary reference resources

LESSON 4: HOW ARE INVERTEBRATES GROUPED TOGETHER?

This lesson follows on from Lesson 3 and deliberately adopts the same pattern and activities in order for children to explore the classification of invertebrates. By the end of this lesson children have a clear understanding of the differences and similarities between the main groups of invertebrates and have reviewed their understanding of how animals are classified.

National curriculum links:

Describe how living things are classified into broad groups according to common observable characteristics and based on similarities and differences, including micro-organisms, plants and animals; give reasons for classifying animals based on specific characteristics

Learning intention:

To explore the classification of the main groups of invertebrates

Scientific enquiry type:

Grouping and classifying things

Working scientifically links:

Reporting and presenting findings from enquiries, including conclusions, causal relationships and explanations of and degree of trust in results, in oral and written forms such as displays and other presentations

Success criteria:

- I can use appropriate vocabulary to classify animals and group invertebrates.
- I can use observable characteristics to group and classify invertebrates.
- I can apply what I know in order to classify an unknown animal.

EXPLORE:

Ask: *In what ways are invertebrates different from vertebrates? Are all invertebrates the same?*

Show the children slides 8, 9, 10 and 11 from the Vertebrate and invertebrate slideshow (Lesson 3, Slideshow 1) and ask them to make notes of the common characteristics of that particular invertebrate using Who has what? (Resource sheet 1). When they are finished, allow them a few moments to underline the characteristics they think are unique to each type of invertebrate.

Explain to children that their challenges in this lesson are similar to those they did in Lesson 3, but this time they are going to look more carefully at the different types of invertebrates (worms/annelids, molluscs, arthropods, insects and arachnids).

The challenges are differentiated by the level of detail about the different types of invertebrates children are asked to notice and explain. All challenges have the same starting point.

ENQUIRE:

Challenge 1: Children create a set diagram

Using the information they have noted from the slideshow and other secondary sources, ask the children to create a set diagram that shows which characteristics are common to all invertebrates and which characteristics are common to each of the sub-groups they have identified. There is a Set diagram template (Resource sheet 2) available if needed.

They should then be able to put this set diagram alongside the one they produced for vertebrates and draw a boundary around them both to show that they are all groups of animals. Ask if they can add the characteristics that are the same for all animals.

Challenge 2: Children create a classification chart

Using the information they have noted from the slideshow and other secondary sources, ask the children to create a classification chart that shows how invertebrates are part of the animal kingdom and includes which characteristics are common to all invertebrates and which characteristics are common to each of the sub-groups they have identified. They should make sure that the different groups are at the correct levels of the chart.

At this stage the children should be able to link this chart to the one they did in Lesson 3 to make a classification chart for all the main groups of animals.

Challenge 3: Children create a set of Top that cards

Using the information they have noted from the slideshow, secondary sources and the cards from Animal card sort (Lesson 3, Resource sheet 1), the children work together to create a class set of Top that cards for a variety of invertebrates. A template is provided in Top that party game (Lesson 3, Resource sheet 4), but the children may prefer to make their own. Each set needs to include different types of insects, annelids, arachnids, molluscs and so on.

Each Top that card must include a photo, the name of the animal, the classification group it belongs to, two characteristics that are shared by that type of invertebrate and four comparable facts that compare common observable and number-based statistics, such as average length, number of legs, number of segments and so on.

Children devise the rules for a game they can play together that helps them in classifying animals. For example, by swapping cards, the winner is the first to collect all the invertebrates that are insects.

REFLECT AND REVIEW:

Ask some children to show their set diagrams and classification charts, and explain how they created them.

Remind them of Professor G Reen who was interested in plants and let them know that he asked one of his colleagues if they could provide a chart showing how animals are classified. By coincidence, it has just arrived.

Compare this chart with children's ones, noting the levels of classification and the different groups, with the examples of the types of animals that belong in each group.

EVIDENCE OF LEARNING:

Observe children as they are classifying their animals. Are they using the appropriate vocabulary? Are they asking and responding to each other's questions in a scientific way? Are they applying what they already know? Do they use observable characteristics to classify animals? Do they know the characteristics of the different groups of invertebrates? Are they more confident in classifying invertebrates?

THE NATURE LIBRARY

Health and safety:

Ensure you adequately prepare for the weather and follow your school's own risk assessment procedure when using the school grounds to support science work. Refer to Be Safe! section 2 when handling living things. Children (and adults) must wash their hands thoroughly whenever they have handled or examined animals, plants, soil, pond water and so on. When working outdoors, if soap and water are not available, ensure that children's hands are cleaned with wipes or gels before eating or drinking anything.

Key vocabulary:

identify, classify, living things, plants, mosses, ferns, conifers, flowering plants, leaves, animals, vertebrates, fish, amphibians, birds, mammals, reptiles, birds, invertebrates, arachnids, annelids, molluscs, insects

Resources:

Recording materials, such as collection pots, paintbrushes (as tickling sticks to encourage small living things into the collection pots for classification), magnifying glasses, identification keys, whiteboards and pens, clipboards and pencils, digital cameras or iPads, a Google map of the school grounds (optional), identification resources from http://www.naturedetectives.org.uk/download/index/ and http://gatekeepersgaffy.co.uk/?page_id=907

LESSON 5: WHERE DO THINGS FIT?

In this lesson children apply some of the things they have learned in the previous lessons in this module to living things in the school environment. They use their knowledge of classification and their identification skills to create a school log book. By the end of the lesson they are able to plan how to record and present findings to show which groups of living things are found around the school. This lesson could be combined with lessons in the Year 6 Our Changing World module.

Preparation required:

Safe access to outside space is required. You may wish to prepare a Google map of the school grounds for children to map the location of the living things.

National curriculum links:

Describe how living things are classified into broad groups according to common observable characteristics and based on similarities and differences, including plants and animals; give reasons for classifying plants and animals based on specific characteristics

Learning intention:

To apply classification concepts to living things in the school grounds

Scientific enquiry type:

Grouping and classifying things

Working scientifically links:

Recording data and results of increasing complexity using scientific diagrams and labels, classification keys, tables, scatter graphs, and bar and line graphs

Success criteria:

- I can create a classification system for some of the living things in the school grounds.
- I can use scientific vocabulary to classify plants and animals.
- I can explain why I have classified certain things in certain ways.
- I can explain the difference between classification and identification of living things.

EXPLORE:

Take children to your chosen outdoor space within the school grounds and, for the first 10 minutes of the lesson, ask them to work independently to record as many living things as they can within the specified area. Try to give them as little guidance as possible at this point. Observe children carefully during this period, paying close attention to those who drift towards other children or adults in search of reassurance. Identify those who begin to classify and identify what they see based on prior learning and those who only focus on particular things, such as animals as a living thing.

Bring the group back together and ask children to share with a partner what they have discovered. Share any similarities and differences in the living things they have recorded. Children do not need to be able to either classify or identify everything they have noted.

At this stage, ensure the whole group clearly understands that living things include both plants and animals. Make sure that you make explicit links to the work already done in previous lessons on classification of plants and animals, and draw on the work children did in Year 4.

ENQUIRE:

Explain to children that their challenge, as a class, is to create a school log book of living things they can find in the school environment. This is done as three individual challenges which each look at a particular type of living thing. Challenge 1 looks at plants, Challenge 2 looks at vertebrates and Challenge 3 looks at invertebrates. There is no differentiation between the challenges, so make sure that groups contain a mix of children who can work well together to achieve the task. Each group becomes the experts for their particular type of living things.

Explain to children that for each living thing that is included in the book, they need to find out what it is called (identification) and in which group of living things it belongs (classification). You may need to discuss this distinction with them to clarify what is needed.

Key information:

Classification is not the same as identification. During classification the emphasis is on the similarities of objects, which demonstrate that they belong to the same group. Identification focuses on the differences between objects in order to be able to give a specific name to that particular thing. The two processes are linked, but not interchangeable.

Key information:

The number of vertebrates is likely to be low but this does not mean that there are not things to find out. It is also worth reminding children that humans are also living things and vertebrates.

Ask: *In addition to recording the name and the group in which a living thing belongs, what else might you want to report in the school log book?*

Suggestions might include: the number of places it is found, how many are seen in each place, what is the most common living thing in the school environment and the date it was seen.

Children could use the identification guides from the Nature Detectives and Gate Keepers's Gaffy websites to name the organisms and use the classification charts to place the plants and animals into the appropriate groups. It is advisable to set up a base camp where the school log book will be collated. Children can come back to it and choose more resources, do further research and bring back their findings. Encourage their curiosity and enquiry skills by providing a place for them to record any new questions that may arise. For example, they might not be able to identify a certain plant or animal from the resources provided and may want to find out more back in the classroom. Perhaps they want to know why there are very few butterflies in the school ground; this might form the basis of some independent enquiry and link to eco-work your school might already be doing.

Challenge 1: Children survey plants, present their findings and draw simple conclusions

The children survey the different types of plants they can find in order to provide the information agreed on for the school log book. Encourage them to think about what they already know about characteristics of particular types of plants, such as leaf shape, seeds, fruits and growing conditions. They should present their findings in appropriate ways and draw simple conclusions about what they have found and where they found it.

Challenge 2: Children gather information on vertebrates and start to draw conclusions

The children gather information about the vertebrates that can be found in the school grounds. Encourage the children to think about what they already know about classification and identification of vertebrates. Ask them to start to draw conclusions about what they have found out.

Challenge 3: Children investigate types of invertebrates and make simple predictions

The children investigate the type of invertebrates that might live in the school grounds. Ask them to decide how they should collect the information and to make simple predictions about what they might find based on concrete facts, such as *I know I am likely to find arachnids in a log pile because ...* and *I know that they are slower when it is colder*. Encourage the children to decide how to classify and identify what they find, and to explain this every time they return to base camp.

REFLECT AND REVIEW:

Bring all the groups back together and review the log book. Encourage children to reflect on information collected by challenges other than their own.

Ask: *Have you collected all the information you set out to collect? Have you been able to classify and identify all of the living things you found? How might you do it for those you don't know? Which is the most common group of living things you have found?*

Ask them to consider how they could present all the findings in a meaningful way to others. How might they answer some of the questions they have generated during their enquiry? These questions could be recorded on different colour sticky notes or centrally on a large piece of paper. Encourage children to independently follow up these questions during the rest of the module.

EVIDENCE OF LEARNING:

Observe and question children as they are classifying and identifying the living things to make judgments about their developing knowledge of classification and identification skills. Can children find examples of the living things they were asked to find? Can they draw conclusions about the school grounds from what they find? Do their responses include appropriate language, such as scientific vocabulary? Can they justify their answers confidently, such as *I know ... because I also know ...?* Do they understand that classification involves grouping things together that have the same or similar characteristics?

CROSS-CURRICULAR OPPORTUNITIES:

The mapping and orientation elements of this lesson link with Geography.

THE NATURE LIBRARY

LESSON 6: WHAT ELSE IS LIVING BESIDES PLANTS AND ANIMALS?

Key vocabulary:

identify, classify, explain, group, micro-organisms (microbes) small, harmful, beneficial (helpful), bacteria, fungi, protista

Key information:

This module refers to micro-organisms, or microbes, to include the three remaining kingdoms of living things: Fungi (moulds and mushrooms); Protista (or Protocista, which includes single-celled organisms that have a nucleus, such as amoeba); and Monera (single-celled organisms which do not have a nucleus and include bacteria). Children may use words such as 'bugs' or 'germs', but these are not scientific terms and children should be encouraged not to use them when talking scientifically.

Resources:

Microscopes (ideally with greater than x8 or x10 zoom, possibly digital), mushrooms

Health and safety:

Great care should be taken whenever micro-organisms are being used. Health and Safety advice in Be Safe! section 8 should be followed at all times.

LESSON SUMMARY:

In this lesson children are introduced to the idea that plants and animals are only two types of living things and that there are three further kingdoms – fungi, bacteria and protista. Together they are often described as micro-organisms. By the end of this lesson children have identified what micro-organisms are within the broader context of living things.

National curriculum links:

Describe how living things are classified into broad groups according to common observable characteristics and based on similarities and differences, including micro-organisms, plants and animals

Learning intention:

To recognise that micro-organisms are groups of living things and explain what they are

Scientific enquiry type:

Grouping and classifying things

Working scientifically links:

Identifying scientific evidence that has been used to support or refute ideas or arguments

Success criteria:

- I can explain that plants and animals are not the only groups of living things.
- I can explain what micro-organisms are.
- I can begin to organise micro-organisms based on their common observable characteristics.
- I can present my findings to others.

EXPLORE:

Ask: *What do you think is the smallest living thing you can see? Do you think there are things that are even smaller?*

Ask children to make a list of everything they can think of and then to share their list with a partner.

As a class, discuss the ideas children have come up with and show What can you see? (Slideshow 1). You may also find it useful to print out a set of the images (four slides per page) per table group to allow children to explore the images more closely. Ask them what they think the images are of and to discuss in their table groups.

If children do not refer to 'micro-organisms' by this stage, introduce the term now. Explain that the images they can see are all of micro-organisms viewed through a very powerful microscope. Explain that micro-organisms are also living things and are divided into three groups: fungi, bacteria and protista.

ENQUIRE:

Explain to children that they are now going to look more closely at what micro-organisms are. All children should do Challenge 1 and then, if there is time, either Challenge 2 or Challenge 3. The challenges are differentiated by the topic to be investigated.

Challenge 1: Children investigate the nature of micro-organisms and present their findings in a poster

Ask the children to look again at the photographs they have already seen and sort them into bacteria, fungi and protista.

Ask: *In what ways are the organisms in each group the same? In what ways are the groups different?*

Using secondary sources, ask the children to find out how we know micro-organisms are small and what are the characteristics of the three kinds. You could direct them towards http://www.cellsalive.com/howbig.htm to explain visually how small different micro-organisms are. Also use microscopes and mushrooms to allow them to look at fungi (a type of micro-organism) close up. You may wish to give Investigating micro-organisms (Resource sheet 1) directly to the group, which provides some information on how to use microscopes to view these micro-organisms.

Children should present their findings in a poster that is visually stimulating and informs others about the different types of micro-organisms and their characteristics. These could form part of a class display for this module.

Challenge 2: Children investigate how micro-organisms can be helpful and share their findings in a presentation

Ask the children to use secondary sources to find out how micro-organisms can be helpful in our lives. Split the group into pairs and ask one child to focus their research into how micro-organisms are useful in food production (yoghurt and cheese, for example) and the other to focus on how micro-organisms can help us with medicines such as penicillin.

They should come together to share their findings in a presentation such as PowerPoint or Smart notebook. Ask them to restrict their slides to six bullet points each to get them to focus on the key facts. They will be asked to share these in the Reflect and review section.

Challenge 3: Children investigate how micro-organisms can be harmful and share their findings via a method of their choice

Ask the children to use secondary sources to find out how micro-organisms can be harmful to us. Give the following statistic to the group: Almost 99% of bacteria are helpful. Disease is caused by only a few of them. Ask them to investigate which diseases are caused by bacteria and to specifically focus on what we can do to prevent them from causing harm to our health.

They should present their information in any way they feel appropriate, remembering that it needs to be helpful to others. These presentations could form part of a class display or be made into information posters to be displayed around the school.

REFLECT AND REVIEW:

Ask children to share some of their posters on micro-organisms. They should quickly and concisely report back to the rest of the class.

Ask those who did Challenge 2 to explain how micro-organisms can be helpful and those who did Challenge 3 to explain how micro-organisms can be harmful.

Ask: *Do you think it is better to have micro-organisms than not?*

EVIDENCE OF LEARNING:

Talk to children to check they understand their particular area of focus. Their posters and PowerPoint presentations will also indicate things they understand and where they are not sure. Can children explain what micro-organisms are? Do they appreciate how small they can be? Can they group micro-organisms in different ways, such as into healthy or harmful ones? Can they classify and name different types of micro-organism, such as bacteria, fungi and protista?

THE NATURE LIBRARY

Key vocabulary:

plan, do, review, risk, micro-organisms, multiply, colony, colonies, mould

Resources:

Petri dishes (at least four per group), trays (one per group), fresh sliced white bread, stale white sliced bread, brown bread, granary bread, sealable transparent plastic bags, sticky tape, labels, cameras

Health and safety:

The safety code for using micro-organisms must be followed (Be Safe! section 8).

Key information:

As a class, agree a set of principles needed to ensure health and safety:

• sealing all containers

• disposing of containers at the end of the investigation without breaking the seal

• if the group decides to grow yeast, this should NOT be in a sealed container, as the gas needs a chance to escape

• dishes need to be clearly labelled and warning notices put up as necessary

LESSON 7: HOW CAN YOU GROW YOUR OWN MICRO-ORGANISM?

LESSON SUMMARY:

In this lesson children plan and set up an investigation to observe how micro-organisms grow and multiply over time. The results of this investigation need to be recorded over time. Weekly opportunities to observe changes are needed during the enquire stage, ideally over four weeks. Observations need only take 10 minutes, so could be planned around other lessons. The Reflect and review session should be carried out at the end of this period of time.

Preparation required:

Gather together the resources (listed on the left) necessary for growing micro-organisms.

National curriculum links:

Describe how living things are classified into broad groups according to common observable characteristics and based on similarities and differences, including micro-organisms, plants and animals

Learning intention:

To investigate the growth of micro-organisms

Scientific enquiry type:

Observing changes over different periods of time

Working scientifically links:

Planning different types of enquiries to answer questions including recognising and controlling variables where necessary

Success criteria:

• I can plan an investigation to grow micro-organisms carefully, considering health and safety.

• I can closely and systematically observe changes in the growth of micro-organisms.

• I can present my findings to others and give reasons for my conclusions about the growth of micro-organisms.

EXPLORE:

Ask children where they might have seen micro-organisms growing. If they are not sure, rephrase the question to where have they seen mould growing. Remind them that mould is a micro-organism.

Ask: *In what type of conditions do you think moulds grow best? Why might investigating micro-organisms be particularly difficult?*

As they think about their answers, remind children of what they learned in the last lesson about micro-organisms being very small and invisible to the naked eye. In order to see them, we need to grow some that multiply until they are visible with the naked eye.

Remind children about the work children did for Challenge 3 in the previous lesson, when they researched how harmful micro-organisms can be.

ENQUIRE:

Explain to children that they are going to plan and set up investigations to find out in which conditions some micro-organisms grow best. At this stage, the challenges all involve growing their own micro-organisms in controlled conditions. Depending on the resources available to you, you should split these groups into smaller groups (ideally twos or threes) to allow children to be as hands-on with their investigation as possible. The challenges are differentiated by the level of support provided for children and the level of detail in the report they are asked to produce at the end of the investigation.

Challenge 1: Children investigate the effects of micro-organisms on bread and produce a report of their conclusions

The children choose from four different types of bread (they can use all of them if they wish) to find out which one stays freshest for longest. Ask the children to plan what they need, how they are going to set up their investigation, how often they will observe for changes and how they will record what they observe. They should also refer to the health and safety considerations agreed in the first part of the lesson. When you are confident that they have effectively planned their investigation, and can clearly explain what they are changing and what they are keeping the same, allow them to select their equipment and set up their investigation.

The children record the changes they see on the bread and then provide a one-page report to show their results and conclusions.

Challenge 2: Children investigate the effects of micro-organisms on bread, and report and explain their conclusions

Explain to the children that there are four types of bread in the shop and they need to identify which one stays fresher for longer. They need to list what equipment they need, when and how they will observe changes, what variables they should control and what health and safety considerations they need to take account of; they also need to make a prediction based on real life experiences.

When you are confident that they are clear about what variables will change and which will remain the same, to ensure the outcomes can be relied upon, they can select their equipment and set up their investigation.

The children record the changes they see on the bread and then provide a report to show and explain their results, giving reasons for their conclusions.

Challenge 3: Children investigate how to prevent the growth of micro-organisms, produce a visual report and give reasons for their conclusions

Explain to the children that the four types of bread provided will all go mouldy over time and that they need to plan an investigation to find out the best conditions to prevent the growth of mould. They should devise a way of measuring or estimating the amount of mould that grows. Estimating abundance of micro-organisms (Resource sheet 1) is provided to support this challenge to estimate abundance when observing the growth of micro-organism colonies over time and it could also be used in Challenges 1 and 2. The principal behind this could be adapted to other types of observation where there are large numbers involved and it would be unrealistic to expect the children to count individual species or objects.

Ask the children to record the changes they see on the bread and then produce a report that presents a visual to others to show the approximate growth rate of colonies, focusing on abundance as a measure. It should also give reasons for their conclusions.

REFLECT AND REVIEW:

After a suitable period of time to allow observations to take place (at suggested weekly intervals over four weeks, with one week for the set-up), give children time to reflect on what they have found out.

If they have taken photographs, they should make comparisons over time. Drawings would enable them to consider this visually.

They present their findings to others. This could be done by setting up groups that include some children who have done each challenge. Each of these groups could then report back what they think are the conditions in which mould is most likely to grow.

EVIDENCE OF LEARNING:

Listen to children during their planning. Can they plan this investigation systematically? Are they aware that this is not a fair test but an observation over time, where the amount of mould that grows on different types of bread in a certain time is being compared? Do they understand that for the findings to be valid, things such as location, temperature and size of bread must be controlled? When they observe at regular intervals are they able to make accurate observations and carefully record what they see? When they draw conclusions do they refer their presentation of results back to the original question? Do they refer to what they might need to do to improve the reliability of this small-scale investigation?

THE NATURE LIBRARY

Key vocabulary:

historically, grouping, classifying, Aristotle, Carl Linnaeus, kingdom, Animalia, Plantae, Fungi, Protista and Monera (or Prokaryota)

LESSON 8: WAS IT ALWAYS THIS WAY?

LESSON SUMMARY:

In this lesson children explore the history of classification and the scientists involved, including Aristotle and Carl Linnaeus. By the end of the lesson children have considered different ways that people in the past classified living things and how these ways have changed through history.

National curriculum links:

Describe how living things are classified into broad groups according to common observable characteristics and based on similarities and differences, including micro-organisms, plants and animals; give reasons for classifying plants and animals based on specific characteristics

Learning intention:

To recognise that the classification system for living things has changed through history and is still changing

Scientific enquiry type:

Finding things out using a wide range of secondary sources of information

Working scientifically links:

Reporting and presenting findings from enquiries, including conclusions, causal relationships and explanations of and degree of trust in results, in oral and written forms such as displays or other presentations; identifying scientific evidence that has been used to support or refute ideas

Success criteria:

- I can describe to others how people classified living things in the past.
- I can explain the importance of Carl Linnaeus in the way we classify living things.
- I can suggest reasons why the classification systems change over time.

Key information:

The five kingdoms of living things are: Animalia – animal kingdom; Plantae – plant kingdom; Fungi – fungi kingdom; Protista – single-celled organisms with a nucleus (micro-organisms such as slime mould and algae are part of this kingdom); Monera – single-celled organisms without a nucleus (micro-organisms such as bacteria are part of this kingdom).

Key information:

At first glance fungi, especially mushrooms and toadstools, do resemble plants in that they appear to grow in the ground. However, the key difference is that plants can produce their own food by photosynthesis. Fungi cannot produce their own food, but secrete chemicals that break down organic matter in the environment around them (the substrate) and then absorb the nutrients that are released.

EXPLORE:

Remind children of the work they have been doing on the classification of living things.

Ask: *Do you think they have always been classified in this way? Give reasons for your answer.*

Use the think, pair, share strategy for children to share their key ideas. Show the history of classification (Video 1). Ask children to reconsider their ideas on how the classification of living things has changed over time.

ENQUIRE:

Explain to children that they are going to investigate how and why changes in classification have occurred. The challenges are differentiated by the topic that is to be covered and the way in which the findings of children are to be presented to the rest of the class.

Challenge 1: Children represent Animalia in a freeze-frame or mime

Children work together in small groups (no more than two or three) to present a freeze-frame or mime to represent the features of the five types of Animalia.

Ask: *Do you think Linnaeus found it easy to classify the animals in this way? If not, why not?*

During the Reflect and review session, the children present their dramas to the rest of the class. Encourage the children to refer back to written work from previous lessons in this module to clarify characteristics of the different groups. They may wish to use secondary sources to investigate any groups they are less certain of and how Linnaeus worked out his classification system.

Challenge 2: Children create a poster on Plantae and Fungi

Children focus on the Plantae and Fungi kingdoms. They work either alone or in pairs to create a poster showing the differences between the characteristics of each kingdom. Let the children choose their most productive working style, either alone or in pairs. They should draw upon existing knowledge from this module about classification of plants into the different divisions. They need to do further research about the characteristics of the Fungi kingdom.

Ask: *For a long time fungi were classified as plants. Why do you think this might have happened? Why do you think scientists decided that fungi and plants are separate groups of living things? What is the main characteristic that plants have in common but fungi do not?*

Key information

The kingdoms Protista and Monera were only distinguished relatively recently as scientists became able to study them more closely with improved optical microscopes, electron microscopes and other apparatus for monitoring the growth and structure of them. Linnaeus is unlikely to have been aware of these organisms simply because he did not have the apparatus to study them closely.

Key information:

Linnaeus could only base his classification on observable features of organisms, but scientists today are using DNA analysis and sequencing, among other techniques, in order to determine which organisms are closely related. The results indicate that, although the overall system remains sound, some organisms may be in the wrong groups. Of course, this is not helpful if you are looking at a plant or animal in the middle of the jungle, so a classification system based on observable features is still very important.

Challenge 3: Children investigate micro-organisms and present their findings in a way of their choice

Ask the children to focus on the Protista and Monera kingdoms and carry out further research linked to the micro-organisms work they did in Lessons 6 and 7 to identify the key characteristics that define these two groups. They present their findings in a way they feel is suitable. Suggestions include a computer presentation such as comic life or PowerPoint, a poster or a PMI table.

This is very challenging for children of this age. Only present this challenge to the children that you are confident can already explain the key characteristics of the other three more familiar kingdoms presented in Challenges 1 or 2.

Ask: *Do you think Linnaeus knew about the organisms in these two kingdoms of living things? Why do we know more about them now? How do scientists study these organisms?*

REFLECT AND REVIEW:

During this session, ask each group to present their findings to the rest of the class. Focus on the key characteristics they present and as a class create a list for each kingdom. This could form part of the class display.

Refer back to the video on the history of classification.

Ask: *Why do you think the classification of living things has changed? Why do you think the system designed by Linnaeus is so helpful? Do you think the classification of living things will change in the future – why?*

EVIDENCE OF LEARNING:

Listen to children whilst they are engaging in their tasks. Are they able to select and deselect appropriate secondary sources to back up their ideas for their presentation? Are they able to present their findings clearly and succinctly highlighting the differences between their kingdoms or groups within the kingdom (as in Challenge 1)? Do they understand that classification systems are helpful? Do they know that the system has changed over time? Do they recognise the role played by Linnaeus? Can they suggest reasons why the classification system has changed over time and why it might change in the future?

CROSS-CURRICULAR OPPORTUNITIES:

There are links with History in the timeline of historical changes of how scientists classified living things.

THE NATURE LIBRARY

Key vocabulary:

historically, grouping, classifying, Carl Linnaeus, Phillip Miller, John Ray, botany, conventions

Resources:

Samples of seeds from familiar fruits and vegetables such as cucumber, apples, oranges, tomatoes, strawberries, grapes, pumpkin, pomegranates, aubergines, peppers, chillies or butternut squash; these should be presented to children once they have been removed from the fruit or vegetable, washed and dried; pictures of the plants the seeds came from, trays or boards to sort on, labels for the classification groups, knives to cut open the seeds, hand lenses, binocular microscopes, digital cameras

Health and safety:

Be aware of any allergies in your class when handling seeds. Ensure children use knives safely when cutting up seeds. Supervision is needed (see Be Safe! section 12).

Key information:

Scientists may disagree about many things but they have ways of reaching resolutions. The agreements on what is acceptable are usually called conventions so that all scientists understand what has been agreed. Classification of living things has seen many changes since Linnaeus devised his system. Each change has to be agreed internationally before it is implemented. This can take years as arguments are put for and against the change.

LESSON 9: WHAT HAPPENS WHEN SCIENTISTS DISAGREE?

LESSON SUMMARY:

In this lesson children consider how scientists can have different views on the way things may be classified by using an example of how botanists developed different ways of classifying plants. They then try to reach agreement on their own classification systems for seeds. By the end of this lesson children understand that scientists do not always agree, but that they find ways of reaching agreements which may be changed in the light of further evidence.

National curriculum links:

Give reasons for classifying plants and animals based on specific characteristics

Learning intention:

To explore, using the example of plant classification and children's own classification of seeds, how scientists handle disagreements in science

Scientific enquiry type:

Finding things out using a wide range of secondary sources of information

Working scientifically links:

Identifying scientific evidence that has been used to support or refute ideas

Success criteria:

- I can describe ways in which scientists disagreed about the best way to classify plants.
- I can devise a classification system for seeds and explain why I think it works.
- I recognise that scientists can agree things, such as the way to classify plants, but that the agreement may change if they find new evidence to show there is a better way.

EXPLORE:

Ask: *Do you think scientists always agree? What makes you say that?*

Explain to children that they are going to watch a video of three scientists (botanists) discussing the classification of plants. Children watch Classification of plants – a long story! (Video 1). Ask them to listen carefully and note things that the botanists in the video disagreed on. In groups of four, ask children to agree a list of the three most important things the botanists disagreed about.

These groups of four should then make groups of eight and agree on a final list of three points to share with the class.

Ask: *How did the botanists overcome disagreements? How did you overcome disagreements when you were making your list of three? Was this the same as how the botanists solved their disagreements?*

Possible answers might include: they went away and carried out more research to prove their point, or, they went to another organisation to persuade them they were right.

ENQUIRE:

Explain to children that they are now going to try and reach agreement on a single classification system for seeds.

They should draw upon what they have learned during the module about how to group and classify living things and learn the lessons of history by looking at all observable features, as well as applying what they already know.

Each challenge is differentiated by the number of seeds children have available, the information given as part of the challenge and the particular equipment available to them.

Challenge 1: Children devise a two-level classification of five or six seed types

Ask the children to work in pairs and present them with a selection of seeds from common fruits and vegetables (see the suggestions in Resources). Limit the range to a maximum of five or six seed types, but provide several examples of each seed. Do not, at this stage, provide the fruit or vegetable from whence it came. Ask the children to examine the seeds using the hand lenses

Key information:
Refer back to Lesson 1 to remind children that a classification system has levels and groups. It is not just putting things into groups. It would be helpful to have the classification charts on display to remind children of the structure.

provided and devise a two-level classification system for the seeds based on the observable features of the seeds. They can use information they know about the seeds, but only if they can provide evidence to support their claim, such as an extract from a book or information from the internet. They should label the levels and groups of seeds that make up the classification. Record the classification system by laying it out with the labels and photographing it.

Challenge 2: Children devise a two- or three-level classification of seven or eight seed types

Ask the children to work in pairs and present them with a selection of seeds from common fruits and vegetables taken from the suggested list in Resources. Limit the range to seven or eight seed types maximum, but provide several examples of each seed. Do not, at this stage, provide the fruit or vegetable from whence it came. Ask the children to examine the seeds, using the hand lenses provided and ask them to devise a two- or three-level classification system for the seeds based on the observable features of the seeds. They can cut them open and use information they know about the seeds but only if they can provide evidence to support their claim, such as an extract from a book or information from the internet. They should label the levels and groups of seeds that make up the classification and list the main characteristics for each group. Record the classification system by laying it out with the labels and photographing it.

Challenge 3: Children devise a two- or three-level classification system of a wide range of seeds

Ask the children to work in pairs and present them with seeds from common and more unusual fruits and vegetables taken from the suggested list in Resources. Provide the full range of samples that are available. Provide photographs (or samples) of the fruit or vegetable from whence the seeds came. Ask the children to examine the seeds, using the hand lenses and binocular microscopes provided and ask them to devise a two- or three-level classification system for the seeds based on the observable features of the seeds. They can cut them open and use information they know about the seeds, but only if they can provide evidence to support their claim, such as an extract from a book or information from the internet. They may also take digital photographs of the seeds and load them onto a computer and examine them on screen, zooming in and out to change the magnification. They should label the levels and groups of seeds that make up the classification and list the main characteristics for each group. Record the classification system by laying it out with the labels and photographing it.

REFLECT AND REVIEW:

This section of the lesson has two parts. The first is to set up groups of up to six children that include some who have done Challenge 1, some Challenge 2 and some Challenge 3. Ask them to share the classification systems they have developed.

Ask: *In what ways are the systems the same? Have you used the same characteristics to make groups? In what ways are they different?*

Challenge them to come to an agreement about which they think is the best classification system and to give reasons why.

The second part of this section is for groups to present their 'best classification system' to the class.

Invite children to think about the way in which they worked to produce their classification systems.

Ask: *Was it easy to group the seeds? Did you all agree on the best way? How did you get over your disagreements? Did having more equipment (as in Challenge 3) make it easier or harder to reach an agreement? Did having more information change the classification groups? If so, how? What else might you do to improve the classification systems further? Do you think there is only one way of classifying things?*

EVIDENCE OF LEARNING:

Observe and listen carefully to children when they are classifying their seeds. Are they applying existing knowledge? Are they using common observable characteristics to classify the seeds? Are they able to adapt argument when they learn new knowledge? Do they recognise that decisions are based on evidence from different sources? Were they able to use evidence to resolve disagreements? Did they understand that some disagreements cannot be resolved without further evidence?

THE NATURE LIBRARY

Key vocabulary:

historically, grouping, classifying, characteristics, genus and species

LESSON 10: WHAT SHOULD WE CALL IT?

LESSON SUMMARY:

In this lesson children apply their learning from the module in an unfamiliar context. By the end of this lesson children have carried out a focused activity, using evidence and their own knowledge to classify and name an unknown organism.

Preparation required:

Feature cards for plants (Resource sheet 1), Feature cards for vertebrates (Resource sheet 2) and Feature cards for invertebrates (Resource sheet 3) need printing and cutting up before the lesson. One sheet is enough for four children.

National curriculum links:

Give reasons for classifying plants and animals based on specific characteristics

Learning intention:

To use evidence and apply existing knowledge to classify and name an unknown animal or plant

Scientific enquiry type:

Grouping and classifying things

Working scientifically links:

Presenting findings from enquiries in oral and written forms such as displays or other presentations

Success criteria:

- I can use evidence and my existing knowledge to classify and name an unknown organism.
- I can explain why I have classified and grouped my living thing in a certain way.
- I can present my findings as part of a museum display.

Key information:

The convention for naming organisms is that it should have two parts made up of the genus to which it belongs (a bit like a surname) and the species name (a bit like a given name). The genus is determined by matching the organism to the appropriate group as agreed by the classification system being used. The species name is given specifically to the organism in question so often is named after the person who found it. Together the genus and species provide a unique name for the organism.

EXPLORE:

Ask: *How would you go about classifying an unknown plant or animal? What would you to do? What sort of information would you need?*

Show The explorer's ship (Slideshow 1) to set the scene. Explain to children that they are going to become museum curators and will need to identify and classify the new living things found on the ship for display in their museum.

Ask children to look back through their books and photos and to have discussions with peers to remind themselves of the different ways they have classified plants (flowering plants, mosses, ferns, fruit bearing, deciduous, coniferous or by leaf shape), vertebrates (mammals, fish, birds, reptiles and amphibians) and invertebrates (arachnids, annelids, molluscs and insects). Remind children that the purpose of revisiting these is that they will need to apply their knowledge in order to classify and place living things in the museum accurately.

The challenges differ according to the group of organisms that each group investigates and the detail they need to include for the museum display.

ENQUIRE:

Explain to children that they are going to collect various features of their living thing, and that these will act as clues to which group of living things it belongs to. They should conclude with a name for their created living thing and be prepared to share with another child.

Challenge 1: Children classify and identify a plant

Ask the children to work independently, with adult support if needed, to select one feature card from each of the six piles (shuffled and face down): plant type, leaf shape, leaf veins, leaf surface, fruit type and growing conditions (Resource sheet 1).

Explain to them that they need to design a display card for the plant in question. This should present images of what the plant would have looked like in its native habitat. Ask the children to stick the feature around the edge of a piece of paper and to use the space in the middle to illustrate and write about their plant. Encourage them to focus on things like the leaf detail, seasonal differences and its habitat, and to add as much appropriate scientific vocabulary as possible. They should also give the plant a name.

Ask: *Does it resemble any plants we can find today? Will this help to classify your plant?*

Challenge 2: Children classify and identify a vertebrate

Ask the children to independently select a set of vertebrate feature cards (Resource sheet 2). They should select one card from four of the six piles (shuffled and face down): reproduction, respiratory system, habitat, body covering, limbs and diet. Reassure them they will get more information later.

Explain to them that they are to design a display card that will go up in the museum to explain what is known about the animal in question. They should also give the animal a name.

Explain to them that they only have limited information to work with to identify and classify this vertebrate. They should stick their feature cards around the outside of a sheet of plain paper and use the central space to draw and label as much as they can infer from the feature cards about the vertebrate's appearance, habitat and diet.

Encourage the children to look beyond the obvious and infer information based on what they know. Challenge them to apply what they know about vertebrates but, by the very nature of this activity, you will probably end up with some vertebrates that don't follow the 'rules'.

When you feel the time is right, offer the remaining two feature piles and let the children choose another two cards. Give them a moment to consider what assumptions may need to change now that they have new information. Do not leave it too late to offer the extra information, as it may be disheartening to some children who may feel their existing work was in vain. This feeling is worth exploring, however, as it is a very real emotion felt by scientists in all fields and how they overcome this to continue their research is an important acknowledgement.

Challenge 3: Children classify and identify an invertebrate

Ask the children to work independently on an invertebrate by selecting one card from each of the five piles (shuffled and face down): body parts, legs, wings, body covering and reproduction. (Resource sheet 3)

Explain to them that they are to design a display card that will go up in the museum to explain what is known about the animal in question. They should also give the animal a name.

They should stick their cards around the edge of a piece of plain paper and use the central space to record images and words that would describe their invertebrate. They should include its habitat, its food source and as much detail as possible.

At a suitable time, ask them to select a wild card. They should then think carefully and adapt what they already know to incorporate this new fact. After alterations have been made, encourage the children to reflect on how they felt when they were given an extra fact. What did they have to do? Did this undo existing knowledge?

After a suitable period of time, offer them a further wild card and ask them to repeat this process.

REFLECT AND REVIEW:

Ask children to pair up with another child from a different challenge to share the key characteristics of their living thing. They should tell their partner how they know this and, where possible, refer to a lesson or experience that helped them learn or understand this idea. Their partner should offer a 'praise point' (something they think their partner did that demonstrated good understanding of scientific concepts) and a 'pick-up point' (something their partner could have thought about or added that would demonstrate deeper understanding).

When completed, the display cards could be displayed in the 'class museum'.

EVIDENCE OF LEARNING:

While children are classifying their living things, talk to them and ask probing questions where answers will demonstrate they have understood concepts such as *It can't be an amphibian because it only breathes with lungs.* Do children link the traits to the correct types of animal or plant? Are they able to justify any assumptions they make about the type of organism they had? Are they able to use the naming convention for their animal or plant? Do they understand how new information can change their ideas?

THE NATURE LIBRARY

Resources:
Access needed to suitable ICT tools to publish the book

Key vocabulary:
plants vocabulary, vertebrates (five main groups), invertebrates (four groups, already discussed), classify, identify, explain, describe

E

ENRICHMENT LESSON 1: CAN YOU MAKE A NATURE GUIDEBOOK FOR YOUR SCHOOL?

LESSON SUMMARY:

In this lesson children apply their knowledge from the module to a real life context by looking at living things that they have in their home and their school environment. This activity would work very well linked to Year 6 Our Changing World as part of an ongoing piece of work. By the end of this lesson children have produced a book to share the features of the living things with others.

National curriculum links:
Describe how living things are classified into broad groups according to common observable characteristics based on similarities and differences; give reasons for classifying plants and animals based on specific characteristics

Learning intention:
To consolidate and apply knowledge of the classification of living things to a real life context

Scientific enquiry type:
Grouping and classifying things

Working scientifically links:
Reporting and presenting findings from enquiries including conclusions, causal relationships and explanations of and degree of trust in results, in oral and written forms such as displays or other presentations

Success criteria:
- I can produce a guidebook that explains the key features of the living things in my school grounds.

EXPLORE:

Ask children to recall some of the work from earlier in the module when they identified plants, vertebrates and invertebrates that are in the school grounds.

Ask: *What types of plants/animals did you find in the school grounds? Where did the different groups congregate? How did you classify the living things you found? Which was the most common type of animal/plant?*

Children may wish to spend time reflecting on the relevant sessions.

Key information:
This could be linked to an enterprise project to raise funds for the school or for charity by selling copies of the book to parents and friends of the school. It could also form part of an eco-activity.

ENQUIRE:

Explain to children that they are going to collectively write a book to help visitors and other classes in the school to classify and identify the living things in your local area.

The challenges are differentiated by the tasks that children are asked to carry out.

Challenge 1: Children work in a group to create the plants chapter of the guidebook

They should divide up the work carefully and agree what they wish to include in their chapter. Direct them to include photographs, names of plants and growing conditions, and to create a map of the school grounds that illustrates the location of the plants. Ask them to use appropriate scientific vocabulary and remind them of the work they have already done about this and to draw on it where necessary, rather than repeating work.

They may decide to make fact cards or a template for each plant to ensure they produce information that looks the same visually.

Challenge 2: Children work in a group to create the invertebrates chapter of the guidebook

They should produce information that includes where the invertebrates were found, photos (if taken) or images from secondary sources, abundance, and any other information they feel is relevant to the guidebook.

Suggest that they organise their chapter into four smaller sections to ensure they have included the main invertebrate types investigated so far.

Challenge 3: Children work as a team to create the vertebrates chapter of the book and to act as publishers

They should refer back to what they already know about the school grounds from previous lessons. It is likely that this is the smallest chapter in the book and may be limited to birds and fish. They should seek to include photos, names, location, links to a map, habitat information and a section on the other vertebrate groups and why they are not present in their school grounds. Reptiles are the most likely group to be absent from most schools.

They should also coordinate the publication of this book, including the presentation style, collating everything, and drafting and re-drafting copy produced by children doing the other challenges.

REFLECT AND REVIEW:

Ask children to reflect on the process of producing a book. Ask them to consider how this relates to the work process of scientists and science communicators. In small groups, each with a copy of the book, ask them to consider ways in which this book will help younger pupils, visitors and Year 6 next year when they do this module. Share their ideas in a class group.

EVIDENCE OF LEARNING:

Observe children while they are working on their chapters. How do they use their existing knowledge? Can they identify what they do know and seek out the information they do not have? Do they use scientific vocabulary confidently and accurately? Are they able to transfer their knowledge of the structure of a book (contents, chapters, index) from their work in English and apply it to this project?

THE NATURE LIBRARY

Key vocabulary:

classify, identify, explain, describe, extinct

 E

ENRICHMENT LESSON 2: WHAT HAPPENS WHEN THE LAST ONE LEAVES?

LESSON SUMMARY:

In this lesson children apply their classification skills to extinct or nearly extinct things. This lesson links well with Module 4, Lesson 7, Why do living things become extinct? By the end of this lesson children have considered what extinction is, identified living things at risk of extinction and applied classification skills to extinct animals.

National curriculum links:

Describe how living things are classified into broad groups according to common observable characteristics based on similarities and differences; give reasons for classifying animals based on specific characteristics

Learning intention:

To apply classification skills to build knowledge of extinct or nearly extinct living things

Scientific enquiry type:

Grouping and classifying things

Working scientifically links:

Identifying scientific evidence that has been used to support or refute ideas or arguments

Success criteria:

- I can classify and identify extinct animals.
- I can collate and interpret evidence to increase my knowledge of extinct organisms.

EXPLORE:

Ask children to talk, pair, share their understanding of the word 'extinct'. Collate these together to agree a common understanding. These may well include some examples from children such as dinosaurs and the dodo.

You may wish to take this further by using secondary sources to find more examples.

EXPLORE:

Explain to children that they are going to research various extinct animals and those at risk of extinction, and apply their knowledge of classification to this.

The challenges are differentiated by the task that is set and the type and content of presentation required.

Challenge 1: Children research extinct animals and present their findings as a poster

Ask the children to independently research and present their findings as a poster. They should use Extinct animals: poster ideas (Resource sheet 1) as a starting point. It contains images of familiar animals that are extinct, such as the mammoth and the dodo. They should think about what they already know and classify the list.

Answers: All the examples are vertebrates. Aurochs – mammal, mammoth – mammal, mastodon – mammal, dodo – bird and moa – bird.

They should use secondary sources and the prompts at the top of Extinct animals: poster ideas (Resource sheet 1) for suggestions of what to include in their poster about one of the animals on the list. Encourage literacy skills in skimming and scanning secondary sources and using key information in their own words. They may wish to produce their poster electronically using an appropriate software package.

Challenge 2: Children produce a fact file about four extinct animals

Ask the children to work independently to use secondary sources to produce a fact file about four extinct animals. They should start with Extinct animals: fact file ideas (Resource sheet 2), which will give them ideas for animals they may wish to research. They should start by classifying all the animals on the list with a partner to check before choosing a fourth that they wish to find out more about.

Answers: golden toad – (V) amphibian, Zanzibar leopard – (V) mammal, Po'ouli – (V) bird, madeira large white – (I) insect, tecopa pupfish – (V) fish, Pyrenean ibex – (V) mammal, west African black rhinoceros – (V) mammal, Javan tiger – (V) mammal, spix's macaw – (V) bird, round island burrowing boa – (V) reptile, dutch alcon blue – (I) insect. (V = vertebrate; I = invertebrate)

The animals on this list are more recently extinct and can offer the children doing this challenge an insight into the lessons that can be learned for animals at risk of extinction that they will hear about in the Reflect and review part of the lesson.

Extinct animals fact file (Resource sheet 3) offers the children a writing template to present their fact files. You may wish to offer them a more open-ended approach to presenting than this, but be aware that time taken on presentation of the facts may detract from the research and fact extraction element of the lesson.

Challenge 3: Children research animals at risk of extinction, choose how to present their results and select an animal for further research

Ask the children to work independently to research animals at risk of extinction. They should use Animals at risk of extinction (Resource sheet 4) as a starting point for their research and begin by classifying the animals based on visual features. They should work with a partner to check these before continuing.

Answers: Iberian lynx – (V) mammal, Saiga antelope – (V) mammal, Sumatran tigerr – (V) mammal, Silky sifafka (type of lemur) – (V) mammal, Vaquita – (V) marine cetacean mammal, Javan Rhino – (V) mammal, Cross River Gorilla – (V) mammal, South China Tiger – (V) mammal, Amur Leopard – (V) mammal, Giant Panda – (V) mammal, African Black Rhinoceros – (V) mammal, Scarlet Macaw – (V) bird, Mountain Gorilla – (V) mammal, Sea Turtle – (V) reptile, Nile crocodile – (V) reptile, Siberian tiger – (V) mammal and Giant otter – (V) mammal (V = vertebrate; I = invertebrate)

They choose a presentation style, such as poster, extended writing or slideshow, and select an animal for further research. They should use the prompts at the top of Animals at risk of extinction (Resource sheet 4) for suggestions of what to include in their presentation.

Some of the children doing this challenge may wish to compare animals of the same type that are at risk; for example, there are three tigers, two gorillas and two rhinos in Research sheet 4 and they may wish to explore if there is any connection. They should be encouraged to do so.

REFLECT AND REVIEW:

Ask children to make groups of three, one from each Challenge, to share their findings. Ask children to make links between what the children doing Challenges 1 and 2 have found out about animals that are already extinct and animals that are close to extinction.

EVIDENCE OF LEARNING:

Observe children while they are working on their research. Do they use their existing knowledge of classification? Is it secure? Can they identify what they already know and seek out what they do not?

Key information:

Some children may already be aware of preservation organisations and charities or may have discovered some through their research in this lesson. Be aware that there is some very emotive material when researching this topic and lots of adverts asking for money. It may be that your class feel very strongly about this issue and may wish to pursue it further as a class charity over time.

BODY PUMP

INTRODUCTION

This module builds on learning about the human body from Key Stage 1, when they learned that humans and other animals need water, food and air in order to survive, and also during lower Key Stage 2, when they investigated the muscular, skeletal and digestive systems. In this module children learn about the human circulatory system and how it enables their bodies to function. They find out about the main parts of the circulatory system: the heart, blood vessels (arteries, veins and capillaries) and blood, and how these work together to deliver oxygen and nutrients to every part of the body. They will find out how the heart works, the main components of blood and the function of the different types of blood vessels. They will also learn about how water is transported through the body and develop their understanding of the importance of water to human health.

This module links closely with Module 3, Body Health, in which children find out how to keep their bodies healthy and about the impact of diet, exercise, drugs and lifestyle.

When working scientifically during this module, children will use secondary sources of information with increasing independence in order to find answers to questions about the functions of different parts of the circulatory system that they cannot investigate first hand. This should involve them using quality non-fiction books, web-based material and health education publications. They could also question local medical experts. Children will carry out and illustrate a practical activity in which they make some 'blood soup', and, in a drama activity, they will model the transport of blood and gases around the body. Some teachers may wish to show children the different parts of a sheep's or pig's heart, which can easily be acquired from a butcher. These can be dissected using scissors to make a memorable demonstration lesson. You can find instructions on the CLEAPSS website http://www.cleapss.org.uk/attachments/article/0/Looking%20at%20(Dissecting)%20hearts%20in%20primary%20schools.pdf?Conferences/ASE%202014/. Children will report and present findings from their enquiries in a variety of ways, both orally and in written forms including labelling diagrams, drawing conclusions, identifying causal relationships and explaining their thinking.

Key vocabulary:

aorta, artery, atrium, blood, blood vessel, body temperature, capillaries, carbon dioxide, cells, chamber, chest cavity, circulation, circulatory system, deoxygenated blood, digestive system, digestive tract, health, heart, heart valves, humans, hydration, lubricant, lungs, muscular system, nutrients, nutrition, oxygen, oxygenated blood, plasma, platelets, pump, red blood cell, skeletal, system, transport, valve, vein, vena cava, ventricle, vessel, waste, waste gases, white blood cells

FACT FILE:

The Mammalian Circulatory System

Blood components

- Plasma: A relatively clear, yellow-tinted water containing sugar, fat, protein and salt solution, which carries the red cells, white cells, and platelets. Normally, 55% of the blood's volume is made up of plasma. It is likely that children will be surprised by the colour of plasma, and it should be pointed out that it is the red blood cells that turn it red.

- Platelets: Cell fragments that work with blood clotting chemicals at the site of wounds by sticking to the walls of blood vessels, thereby plugging the gap.

- Red blood cells: Relatively large microscopic cells that normally make up 40-50% of the total blood volume. They transport oxygen from the lungs to the body's living tissues and carry away carbon dioxide.

- White blood cells: There are different types of white blood cells that exist in variable numbers but that collectively make up a very small part of blood's volume – normally only about 1% in healthy people.

Heart

The heart is a very strong muscle that pumps blood around the body. It is made up of four chambers – two upper and two lower. Blood enters the upper chambers which squeeze and push the blood into the lower chambers. Here it is squeezed and pushed out of the heart.

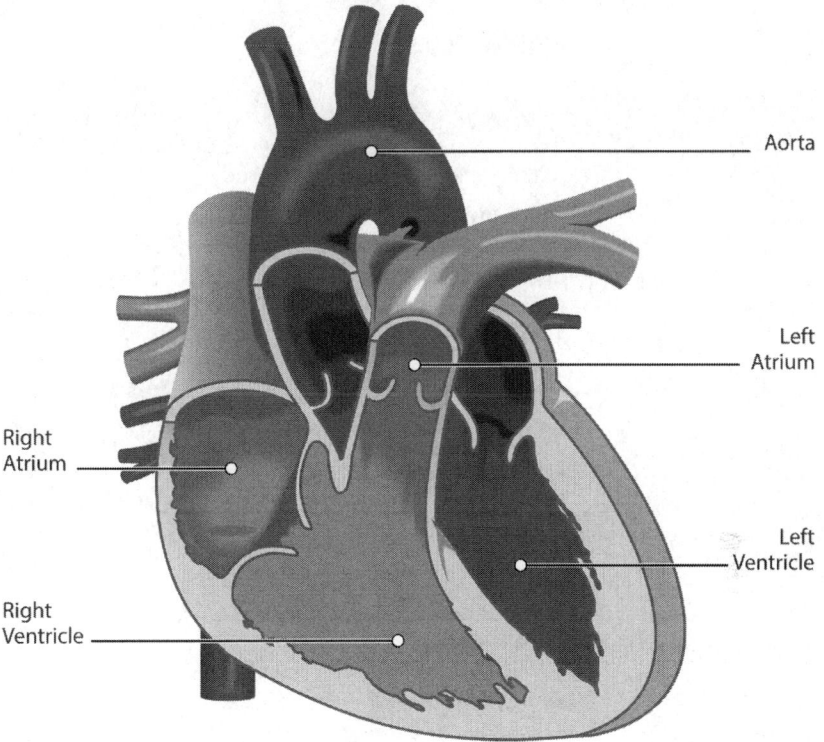

Aorta

Left Atrium

Right Atrium

Left Ventricle

Right Ventricle

Blood vessels

These are the tubes that carry blood around the body. There are three main types:

• Veins carry blood back to the heart. In the veins that collect blood from the body and return it to the heart the blood is deoxygenated. However, the blood in the pulmonary vein, which carries blood back to the heart from the lungs. is oxygenated.

• Arteries carry blood away from the heart. Arteries that transport blood around the body carry oxygenated blood. However, the pulmonary artery, which transports blood from the heart to the lungs, carries deoxygenated blood.

• Capillaries are the small (in some cases very small) blood vessels that carry blood through the various tissues of the body, taking oxygen to the cells and taking away carbon dioxide. In effect they form the links between arteries and veins.

Common misconceptions:

Children may think that heart is love-heart shaped and in the left hand side of the chest. The heart is actually roughly the size and shape of a person's clenched fist. It is located in centre of the chest but 'leans' slightly to left.

Children may think that blood in our veins is blue. In fact, all human blood is red, ranging from bright red when oxygenated to dark red when not. It owes its colour to haemoglobin. Blood is never blue, but veins appear blue because light is diffused by the skin. Red and blue colours are typically used to show oxygenated and deoxygenated blood in scientific diagrams of the human circulatory system.

Children may think that air tubes connect the lungs to the heart. The process of the transfer of oxygen gas from the air into liquid blood is actually more complex. After a breath of air is inhaled it ends up in air sacs (alveoli) in the lungs where it dissolves into the blood across capilliaries. At the same time carbon dioxide leaves the blood and enters the alveoli, ready to be exhaled from the body.

Big Cat book links

| **Extreme Sports** Adrian Bradbury
978-0-00-733632-6
Band 14 Ruby | Find out about the dangers people face in extreme sports |

BODY PUMP

 ## LESSON 1: WHAT DOES MY CIRCULATORY SYSTEM DO?

LESSON SUMMARY:

In this lesson children begin their investigations of the human circulatory system, first revising knowledge of the digestive, muscular and skeletal systems from Year 4. By the end of the lesson children are able to name the parts of the circulatory system and describe how it transports blood containing oxygen around the body.

Preparation:

Make three body part tabards by attaching the images of the heart, lungs and body (Resource sheet 1) to tabard style sports bibs prior to lesson. Stick sheets of red and blue paper back to back. You will need a large space indoors or outside to draw or mark a giant figure of eight (big enough for up to 10 children to be standing on spaced apart) to model the circulatory system.

National curriculum links:

Identify and name the main parts of the human circulatory system and describe the functions of the heart, blood vessels and blood

Learning intention:

To describe how the human circulatory system works

Scientific enquiry type:

Finding things out using a wide range of secondary sources of information

Working scientifically links:

Recording data and results of increasing complexity using scientific diagrams and labels, classification keys, tables, scatter graphs, and bar and line graphs

Success criteria:

- I can identify the human digestive, muscular and skeletal systems.
- I can name the parts of the human circulatory system.
- I can explain how the parts of the circulatory system work together to enable the human body to function.

EXPLORE:

Place large (A3 or larger) print-outs of the skeletal, muscular and digestive systems (Resource sheet 2) at different stations around the room. Ask children to work in small groups to visit each station adding any names (basic or scientific) or functions to these posters.

Use Interactive 1, Body systems, to ensure that children can recognise the skeletal, muscular and digestive systems, simply describe their functions and name some of the main parts of each. Ask children to agree a definition for a 'system'.

ENQUIRE:

Explain to children that they are now going to learn about the circulatory system that transports blood containing oxygen, nutrients and water around the body to keep us alive.

Explain to children that there are only three parts to the circulatory system: the **blood**, which circulates or travels round the whole body and which escapes when they cut themselves, the **blood vessels** that carry the blood, which they may be able to see where their skin is thin, such as inside their wrists, and the **heart**, which is the muscle that pumps the blood round the body. Show children the tabard with the picture of the heart.

Ask: *Can you feel your heart pumping? Can you see any blood vessels? What happened to the blood last time you cut yourself?*

Tell children they are going to investigate how the three parts work together to keep them alive. Explain to them that do this they need to know about another important body part – the lungs. Ask them to put their hands on the top of their ribs and breathe in deeply.

Ask: *Can you feel your lungs expand?*

Show them the tabard with the picture of the lungs. Tell them that there is oxygen in the air that they breathe in and that every part of their body needs oxygen to function.

Explain to them that they are going to act out how the circulatory system works. The challenges are differentiated by the level of detail required in the diagram and the extra information they are required to find out about the way the parts of the system work together.

Key vocabulary:

heart, blood vessels, veins, arteries, blood, system, lungs, circulatory system, skeletal system, muscular system, digestive system, oxygenated blood, deoxygenated blood, nutrients, water

Resources:

Chalk or masking tape, three tabard style sports bibs, bicycle or foot pump, stethoscopes or cardboard tubes, sheets of red and blue paper stuck back to back

Key information:

This lesson focuses on how the circulatory system transports oxygen to all parts of the body. In later lessons children learn about the circulatory system's role in transporting nutrients and water, also essential for human life.

Key information:

A system is a set of connected organs or parts forming a whole, or a set of things working together as parts of an interconnecting network.

Draw a very large figure of eight on the floor. Explain to children that the lines represent the main blood vessels which carry blood around their whole bodies. Give one child the heart tabard to wear and ask them to stand on the middle of the eight. Give them a foot or bicycle pump. Ask another child to wear the lungs tabard and to stand at the top of the eight. A third child should wear a tabard with the whole body image and stand at the bottom of the eight.

Ask someone to volunteer to be some blood circulating the body. Ask the heart to start pumping, and then the blood should start from the heart and walk up the left side of the figure eight to the lungs, past the lungs and back down the right side to the heart, then from the heart to the body along the right side of the eight and then back to the heart. The heart must keep pumping all the time.

Once the 'blood' has started circulating, ask other children to join the system to show that blood is circulating in all parts of the body all the time. Other children should check that all the 'blood' is travelling in the right direction and that the heart keeps on pumping.

Challenge 1: Children complete a diagram of the human circulatory system and use secondary sources to add more information about what happens to the blood

Give the children the outline of the figure-of-eight circulation system (Resource sheet 3) to complete.

Ask: *Can you show which direction the blood travels around the human body? Can you find out what happens to the blood in the lungs? What is the difference between the blood on the right side of your diagram and the left? Can you find out the names of any more parts of the circulatory system and add them to your diagram?*

Challenge 2: Children complete a diagram of the human circulatory system and use secondary sources to add more about what happens to the blood in the lungs and rest of the body, and the function of veins and arteries

Give the children the outline of the figure-of-eight circulation system (Resource sheet 4) to complete.

Ask: *Can you find out what happens to the blood in the lungs and the rest of the body? What is the difference between the blood on the right side of your diagram and the left? What are veins and arteries? What do you think might happen if there were a blockage in a vein or artery? Can you find out the names of any more parts of the circulatory system and add them to your diagram?*

Challenge 3: Children create a labelled diagram of the circulation system

Ask the children to draw their own diagram of the figure-of-eight circulatory system, labelling as many parts as they can and showing the direction the blood circulates.

Ask: *What are the key features of each part? How do the parts work with each other to move the blood around the body? What might happen to the circulatory system if one of the parts were not functioning properly?*

REFLECT AND REVIEW:

Bring the class back to the large figure-of-eight model and choose children to represent the heart (with pump), lungs and body, and to find their place on the model. Choose one child to be the blood and give them one of the red and blue sheets of paper to hold. Explain to them that the red side is blood with oxygen in it and the blue side is deoxygenated blood. Once the heart starts pumping ask the 'blood' to walk from the left side of the heart to the lungs.

Ask: *What type of blood are you now? What colour should be showing?*

When they reach the lungs:

Ask: *What happens to the blood here? What colour should be showing?*

Choose a child to direct the 'blood' to complete an entire circuit, changing the card to show whether the blood is oxygenated or deoxygenated at each stage. Once everyone agrees that the 'blood' is accurately modelling what goes on at each stage of the circulatory system, give other children blood cards to hold and ask them to join the system, turning the cards over as they walk from part to part.

Now ask children to return to their diagrams and use red and blue pens to colour the arrows they have drawn (to show the blood vessels) to show which type of blood is in each part.

EVIDENCE OF LEARNING:

Look at children's annotations to the body systems posters and their labelled diagrams of the circulatory system, and observe as they enact the model of the circulatory system. Do they explain that a system is a set of different parts working together? Can they name the main systems of the human body? Can they name the main parts of the circulatory system? Are the parts correctly positioned and labelled on their diagrams? Do they explain that blood transports oxygen around the body? Do they colour the blood vessels correctly?

Key information:

Make sure that children understand that the blue and red in their diagrams and those in books represent the different types of blood but that blood is never blue. Scientists use different colours to show the difference clearly in diagrams.

BODY PUMP

 LESSON 2: WHAT IS A HEART AND WHAT DOES IT DO?

LESSON SUMMARY:

In this lesson children make a model of the heart to illustrate how the different parts fit and work together. By the end of the lesson children have an understanding of the heart and its main functions in the circulatory system.

National curriculum links:

Identify and name the main parts of the human circulatory system and describe the functions of the heart, blood vessels and blood

Learning intention:

To investigate and describe the main functions of the heart

Scientific enquiry type:

Finding things out using a wide range of secondary sources of information

Working scientifically links:

Reporting and presenting findings from enquiries, including conclusions, causal relationships and explanations of and degree of trust in results, in oral and written forms such as displays and other presentations

Success criteria:

- I can explain the main functions of the heart.
- I can represent the main parts of the heart as a 3D model.

Key vocabulary:

aorta, artery, atrium, blood, capillaries, chamber, circulation, heart, heart valves, vein, ventricle, vessel, pump, oxygen, lungs, rest of body, chest cavity, circulatory system

Resources:

Different coloured modelling clay (enough for children to work in pairs), scissors, base boards, digital camera(s) and cocktail sticks (optional), access to secondary sources of information about the heart, children's labelled diagrams of circulatory system from Lesson 1

Health and safety:

Care should be taken if you decide to use cocktail sticks for labelling the heart in the Explore section.

Key information:

You may wish to demonstrate how a mammal heart works by using scissors to dissect a lamb or pig's heart. This is perfectly possible in a primary classroom, as long as appropriate health and safely guidance is followed.

EXPLORE:

Ask children to look at their diagrams from Lesson 1. Ask them to write a sentence on their own diagram to remind them what they learned about the circulatory system. They should compare their sentence with a partner.

Explain to children that in this lesson they are going to learn about the different parts of the heart. Make sure that children remember where their heart is and that it pumps blood around the whole body, never stopping.

ENQUIRE:

Show children Animation 1, The heart in action, which explains in detail what happens when blood passes through the heart.

Ask: *Can you feel your heart beating? What is it doing?*

Remind children that the arteries (shown in red) carry blood from the heart to the rest of the body and the veins (shown in blue) carry blood to the heart. Make sure that children recognise that the heart is made of four chambers: two upper and two lower.

Explain to children that they are going to work in pairs to make a labelled 3D model of a human heart from modelling clay. The intention of the challenges is for children to have the chance to hold and feel the different parts of the heart to enable them to have a deeper understanding of how it works and fits together than a drawing and labelling activity would provide. The challenges are differentiated by the level of support given to children to help them plan, make and label their model, and the detail they are required to include.

Challenge 1: In pairs, children make and label a model of a heart, with adult support where necessary

Explain to the children that they are going to make and label a model of a heart, and you will help them if necessary. Provide the children with the Diagram of a heart (Slideshow 1), vocabulary cards plus definitions (Resource sheet 1), modelling clay and a base board to make their model on.

Ask: *Can you make the four chambers? What size and shape are they? Can you label them correctly? Where does the blood enter the heart? Where does it leave? Can you label those parts? Can you write a title to go with your heart that states what its function is?*

Challenge 2: In pairs, children make and label a model of a heart

Explain to the children that they are going to make and label a model of a heart. Provide the children with the Diagram of a heart (Slideshow 1), vocabulary cards plus definitions (Resource sheet 1), modelling clay and a base board to make their model on.

Ask: *How many parts do you need to make? Which parts will you need to make first? How will you join them together? How will you show the difference between the veins and arteries? How many parts can you label? Can you write a key to go with your heart to show how it functions?*

Challenge 3: In pairs, children make and label a model of a heart, defining the function of each part

Explain to the children that they are going to make and label a model of a heart and say what each part does. Provide the children with the Diagram of a heart (Slideshow 1), vocabulary cards (Resource sheet 2), modelling clay and a base board to make their model on. They must use secondary sources such as books, posters or the internet to write a definition in their own words for each label.

Ask: *Have you made all the main parts? Have you named them and described what they do? How does your model show the way that blood moves in the heart?*

REFLECT AND REVIEW:

When all the models are complete, ask children to swap their model with another pair. Remind them that one of the success criteria for the lesson was to represent the main parts of the heart in a 3D model. Ask them to evaluate how well the heart they are looking at has met these success criteria. Can they see all the main parts? Are they the right size and shape? Have they used different coloured modelling clay to show the difference in the blood in different parts? Are the parts clearly labelled? Is the heart's function clear? Could anything be improved?

EVIDENCE OF LEARNING:

Look at children's models of the heart and listen as they evaluate another model. Have they labelled the parts of the heart? Are their models accurate and to scale? Do they show the direction that the blood is transported? Can they explain that the heart's main job is to pump blood around the body? Do they describe how the heart also pumps blood to and from the lungs where oxygen is absorbed?

Key information:

In Module 3, Body Health, children will investigate pulse rate further.

BODY PUMP

 ## LESSON 3: WHAT IS BLOOD?

LESSON SUMMARY:

In this lesson children pose and answer different types of questions to find out how blood transports oxygen and waste gases round the body. By the end of this lesson children will have presented and evaluated the way they have answered their questions about what blood is and what it does.

Preparation:

Prepare clear containers and blood-like liquid as detailed in Resources.

Key vocabulary:

blood, vessel, artery, vein, valve, red blood cell, plasma, oxygen, carbon dioxide, waste gases

Resources:

A large bucket capable of holding nine pints (or five litres) of red liquid such as food colouring in water, five empty one-litre drink cartons, plastic funnel and jug

Health and safety:

Remind children to be cautious around large volumes of liquid in the classroom at the start of this lesson.

National curriculum links:

Identify and name the main parts of the human circulatory system and describe the functions of the heart, blood vessels and blood

Learning intention:

To pose and answer a range of relevant questions about how blood transports gases round the body

Scientific enquiry type:

Finding things out using a wide range of secondary sources of information

Working scientifically links:

Identifying scientific evidence that has been used to support or refute ideas or arguments

Success criteria:

- I can explain the function of red blood cells.
- I can pose different types of questions to find out more about blood.
- I can identify information sources to answer my questions about blood.
- I can communicate my findings about blood clearly.

EXPLORE:

Show children an empty one-litre drink carton. Ask them to estimate how many of these it would take to hold the blood in an adult's body. Ask them to justify their decision.

Show them the five-litre bucket. Explain that the red liquid in the bucket represents how much blood is in the human body. Ask children to estimate how much they think is in the bucket. Use a funnel and a jug to fill the milk cartons with the liquid. Whose estimate was closest?

Explain to children that you want them to pose questions to help them find out more about how blood transports oxygen and other gases around the body. Explain how questions can be sorted into different types and together generate an example of each type linked to what they have just seen about the amount of blood in an average adult human.

Type A: questions that ask for factual answers

Type B: questions that ask for confirmation of something

Type C: questions that ask for explanations of phenomena or ideas

Type D: questions that speculate or hypothesise about something

Type E: questions that surprise us

Show children Animation 1, Movement of blood, and ask them to listen closely to the information about what the red blood cells do. Ask them to make notes whilst watching and then use Resource sheet 1 to record and classify the different questions they have. What types of questions are they?

ENQUIRE:

Explain to children that you want everyone to answer the question: What do red blood cells do? They can then choose one other question each to answer. They present their findings in the Reflect and review section in whichever way they choose. The aim is to communicate what they have found out about how blood transports gases round the body and other functions, clearly explaining why others should have confidence in their answer.

The challenges are differentiated by the type of additional question the children choose to answer and the support they are given. Encourage children to tackle a question that extends their research skills and prior knowledge about the heart and blood.

Challenge 1: Children work individually to answer their own factual question about blood

Ask the children to answer the question: What do red blood cells do? They then choose another factual question that can be answered by using a secondary information source, find the information they need, record it in their own words and present the answer in an interesting and original way. They may choose to watch the animation again.

Challenge 2: Children work individually to answer their own explanatory or confirmatory question about blood, justifying why they chose the information source

Ask the children to answer the question: What do red blood cells do? They then choose a question that requires confirmation or explanation to answer, decide where to find the information (do they need to ask someone who has personal experience?) and present their answer in an interesting and original way.

Challenge 3: Children work individually to answer their own speculative question about blood, explaining how confident they are in the idea they develop

Ask the children to answer the question: What do red blood cells do? They then choose a question that requires informed speculation to answer, decide what information to base their speculation on and present their answer in an interesting and original way, explaining why they are confident about the idea they have developed.

REFLECT AND REVIEW:

Choose two or three children to present their answers to the question: What do red blood cells do?

Ask other children if they agree or if they can improve the answer at all. Then choose children to present their answers to the different types of questions.

Ask: *Is the answer clear? Are you convinced they are right? How have they convinced you?*

EVIDENCE OF LEARNING:

Observe as children pose their questions, research the answers and present their findings. Do they explain the function of the red blood cells to transport oxygen and waste gases around the body? Do children pose relevant questions that help them find out more about blood? Do they classify the question types correctly? Do they identify appropriate sources of information? Can they select the right information to answer the question? Do they present their findings clearly? Can they evaluate the quality of answers, with reference to the reliability of the evidence source?

BODY PUMP

LESSON 4: WHAT IS IN BLOOD?

LESSON SUMMARY:

In this lesson children make 'blood soup' as an illustrative practical activity to help them find out about how the different parts of blood enable it to carry oxygen, waste gases, nutrients and water, and compile a fact file. By the end of this lesson children are able to list the different parts of blood and describe the function of each part.

Key vocabulary:

blood, platelets, nutrients, digestive tract, white blood cells, red blood cells, plasma

Resources:

Plasma – yellowy liquid such as weak orange squash (one cup per soup); red blood cells – lots of small red jelly sweets or chopped up pieces of a raspberry jelly cube; white blood cells – a few small white marshmallows; platelets – small amounts of white rice; one sealable plastic bag per person or group making blood soup, children's diagrams from Lesson 1

Health and safety:

Remind children to take care to avoid spillages of their blood soup during this lesson. Be aware of cultural or personal sensitivities in this lesson and adapt appropriately.

Key information:

Nutrients are essential substances in food that humans and other animals need to take in in order to stay alive and to grow. They give energy and renew cells. They are absorbed into the blood stream from the digestive system. In Module 3, Body Health, children learn more about different food groups and how they are vital for human health.

National curriculum links:

Identify and name the main parts of the human circulatory system and describe the functions of the heart, blood vessels and blood

Learning intention:

To identify the contents of blood and describe their function

Scientific enquiry type:

Finding things out using a wide range of secondary sources of information

Working scientifically links:

Reporting and presenting findings from enquiries, including conclusions, causal relationships and explanations of and degree of trust in results, in oral and written forms such as displays and other presentations

Success criteria:

- I can name the part of blood that each ingredient of 'blood soup' represents.
- I can explain the function of the different parts of blood.
- I can present my findings about blood in a fact file.

EXPLORE:

Ask children to look at the diagrams of the circulatory system that they made in Lesson 1.

Ask: *What does blood do in this body system?*

Establish that blood transports oxygen, nutrients and water around the body, and that it is pumped by the heart and carried in veins and arteries. Remind children that they learned about how oxygen and waste gases are transported in the blood in Lesson 3. Now they are going to learn about how blood also carries nutrients and water, and clots to stop bleeding.

Ask children what they know about blood donation. Have they ever seen a blood donor van? Do their parents give blood? What is a blood bank? Show Video 1, Why donate? After watching the video, ask children to share with a partner some reasons why it is important that healthy adults donate blood and why people would choose to do it.

ENQUIRE:

Explain to children that they are going to make some 'blood soup' (model blood) using different ingredients to represent the four main parts of blood. Show the first section of the video again and ask children to list the four main parts. Ask them to share their list with a partner: have they listed red blood cells, white blood cells, plasma and platelets?

Explain to children that they have to 'earn' each of the four ingredients to make their blood soup by completing a section of a blood fact file and then visiting their own 'blood bank'. The challenges offer different levels of support to help children find and compile the information they need, plus Challenges 2 and 3 include additional questions about blood donation for children to answer. Children should work independently.

Challenge 1: Children compile a fact file about the four parts of blood and make blood soup

Children use secondary sources of information (www.blood.co.uk) to find out the function of each of the four parts of blood and complete Blood facts: Challenge 1 (Resource sheet 1). As they complete each section, children visit the 'blood bank' to claim the appropriate 'ingredient' to make their blood soup. Help the children to work out how much they need of each ingredient based on their fact file research.

LESSON 4: WHAT IS IN BLOOD?

Key information:

Blood is composed of 55% plasma and platelets, 44% red blood cells and 1% white blood cells. Plasma is a yellowish liquid, mainly made of water, that carries nutrients, hormones, and proteins throughout the body. Platelets are small cells that stop bleeding. Red blood cells distribute oxygen to body tissues and carry waste carbon dioxide back to the lungs. White blood cells fight infections and germs by destroying them or by producing antibodies against them.

Challenge 2: Children compile a fact file about the four parts of blood and the use of blood in an emergency, and make blood soup

The children use secondary sources of information (www.blood.co.uk and www.kidshealth.org/kid/talk/qa/blood.html) to find out the function of each of the four parts of blood and complete Blood facts: Challenge 2 (Resource sheet 2), also including information about how blood is used in emergencies. As they complete each section, children visit the 'blood bank' to claim the appropriate 'ingredient' to make their blood soup, in the right proportions.

Challenge 3: Children compile a fact file about the four parts of blood and investigate clotting

The children use secondary sources of information (www.blood.co.uk, www.kidshealth.org/kid/talk/qa/blood.html and http://anthro.palomar.edu/blood/blood_components.htm) to find out the function of each of the four parts of blood and the importance of clotting. They complete Blood facts: Challenge 3 (Resource sheet 3), also including information about how blood is used in emergencies. As they complete each section, children visit the 'blood bank' to claim the appropriate 'ingredient' to make their blood soup, in the right proportions.

REFLECT AND REVIEW:

Children take their blood soup and fact file to someone else in the room and ask and answer a question about one of the parts of blood in the bag. Ask them to use their fact files to answer the questions. Finally, ask the class to nominate four children to concisely explain what each part of the blood does. These explanations could be displayed as speech bubbles on a class display.

EVIDENCE OF LEARNING:

Look at children's 'blood soup' and fact files. Have they included the four parts in correct proportions? Do they know which ingredient each part represents? Can they explain the function of plasma, platelets, and red and white blood cells? Can they explain how nutrients are absorbed into the blood and transported? Can they explain why it is important for blood to clot? Can they explain the importance of blood donation?

BODY PUMP

 ## LESSON 5: WHAT DO VALVES AND BLOOD VESSELS DO?

LESSON SUMMARY:

In this lesson children use their learning from previous lessons in this module and secondary sources to explore valves and blood vessels. They create concept sentences and maps to present their findings about valves, veins, arteries and capillaries. By the end of this lesson children are able to explain the function of valves and the three different types of blood vessels.

Key vocabulary:

blood, blood vessels, valves, veins, arteries, capillaries, aorta, vena cava, oxygenated blood, deoxygenated blood

Resources:

Access to secondary sources such as reference books and the internet, scissors and glue

Health and safety:

Ensure children take the appropriate precautions when using search tools on the internet.

National curriculum links:

Identify and name the main parts of the human circulatory system and describe the functions of the heart, blood vessels and blood

Learning intention:

To explain the function of valves, veins, arteries and capillaries in the human circulatory system

Scientific enquiry type:

Finding things out using a wide range of secondary sources of information

Working scientifically links:

Recording data and results of increasing complexity using scientific diagrams and labels, classification keys, tables, scatter graphs, and bar and line graphs

Success criteria:

- I can create a concept map to show what I know about heart valves, veins, arteries and capillaries.
- I can explain how valves, veins, arteries and capillaries enable my body to function.

EXPLORE:

Remind children that in Lesson 1 they found out that there are three parts to the human circulatory system.

Ask: *Can you remember what they are and what they do?*

Ask children in pairs to complete Graphic organiser 1 (Resource sheet 1). If necessary, remind them that the three parts are the heart, the blood and the blood vessels. Ask them to compare their answers with another pair. Do they want to change anything? Did they remember that the blood transports oxygen, waste gases, nutrients and water?

ENQUIRE:

Remind children that they have already found out about how the heart works and what is in blood. Today they are going to find out more about the parts of the circulatory system called valves and the different types of blood vessels. Their challenge is to organise the information they find into concept sentences or a concept map of the human circulatory system. The challenges are differentiated by whether children are required to create separate sentences or link several ideas in a concept into a complete map, and the technical level of the words used. All children need to include the words 'valves', 'arteries', 'veins' and 'capillaries' in the concept sentences of maps.

Challenge 1: Children create concept sentences to show what they know about parts of the human circulatory system

Ask the children to cut up the words from the Word bank: Challenge 1 (Resource sheet 2) and use some of them to form a sentence. They should ask someone to look at it to see if they agree. Next they need to look at the words they have left and form as many more sentences as they can. They can add other words of their own and use any twice or more. Encourage them to use all the words. They must use the words with a *. If they do not know what a word means they should look it up in a textbook or online.

Once they have finished, tell them to copy their sentences into their books, underlining the new words they have learned today.

Challenge 2: Children create concept sentences to show what they know about the main parts of the human circulatory system

Ask the children to cut up the Word bank: Challenge 2 (Resource sheet 3) and use some of them to form a sentence. They should ask someone to look at it to see if they agree. Next they need to look at the words they have left and form as many more sentences as they can. They can add other words of their own and use any two or more times. Encourage them to use all the words. They must use the words with a *. If they do not know what a word means they should look it up in a textbook or online.

Once they have finished, ask them to copy their sentences into their books, underlining the new words they have learned today and indicating where they found the information.

Challenge 3: Children create a concept map to show what they know about parts of the human circulatory system and how they link together

Ask the children to cut up the words from the Word bank: Challenge 3 (Resource sheet 4) and arrange them to form a concept map, making links between the nouns with arrows and verbs. They should ask someone to look at it to see if they agree. They can add other nouns of their own and use any two or more times. Encourage them to use all the words. They must use the words with a *. If they do not know what a word means they should look it up in a textbook or online.

Ask: *What word should go at the centre of your concept map? Which words link directly to it? Which do they link to? Which verbs will you need?*

REFLECT AND REVIEW:

Ask children to return to the graphic organiser they completed at the beginning of the lesson. Do they want to change anything now they have found out more? In pairs, ask them to use what they have found out in the challenges to complete a second Graphic organiser 2 (Resource sheet 5) and then compare theirs with another pair. Do they agree what valves, arteries, veins and capillaries do and how they enable the body to function?

EVIDENCE OF LEARNING:

Look at children's concept sentences and maps, and both graphic organisers. Do they know what the main parts of the circulatory system are? Do they use the scientific terms for parts of the circulatory system appropriately? Do they explain that capillaries, arteries and veins are different types of blood vessels? Do they explain that nutrients and gases are transferred through the thin capillary walls? Do they explain that valves control the flow and direction of blood? Can they make sentences or links on a concept map which explain the function that different parts play in the human circulatory system? Can they find information in textbooks and websites to find out about parts that are new to them? Are they able to explain the function of different parts in terms of what would happen if it were missing or failed?

BODY PUMP

LESSON 6: WHAT HAPPENS TO WATER IN OUR BODIES?

LESSON SUMMARY:

In this lesson children learn more about how water is transported through their bodies, building on their knowledge about how the blood transports nutrients and gases. By the end of this lesson children know why water is needed for the body to function.

National curriculum links:

Describe the ways in which nutrients and water are transported within animals, including humans

Learning intention:

To explain how water helps humans' and other animals' bodies to function

Scientific enquiry type:

Finding things out using a wide range of secondary sources of information

Working scientifically links:

Reporting and presenting findings from enquiries, including conclusions, causal relationships and explanations of and degree of trust in results, in oral and written forms such as displays and other presentations

Success criteria:

- I can describe how water is transported within humans.
- I can compare the transport of water in humans with the transport of water in some other animals.
- I can explain the importance of water for human health.

Key information:

Children learned about the transportation of water through the root, stem and leaves in plants in Year 3, and how it is essential for plant health. However, there are significant differences to the human circulatory system. In plants there is no pump and no circulating cells, and liquids, unlike blood, do not move continuously.

Key vocabulary:

water, transport, humans, waste, nutrition, animals, cells, body temperature, hydration, lubricant

Resources:

Access to secondary sources of information about human and other circulatory systems, such as the internet, books, posters or leaflets

Health and safety:

Care should be taken to ensure that any diet-related issues are handled sensitively.

Key information:

Slides 3, 4, 5 and 6 contain further information that can be printed out to provide information for the challenges. Encourage children to refer to these as well as other sources of information.

EXPLORE:

Pose a series of questions for children to discuss, using slide 1 of Slideshow 1, Wonderful water, as a visual.

Ask: *Why do we need water? Is all water the same? What happens if we do not get enough water? What happens if we get too much?*

Encourage children to think, pair and share their ideas.

Show slide 2, which provides some exemplar answers. Ask children to compare their answers with these. Discuss each of the ways that water is used in human bodies to check that children understand the descriptions.

ENQUIRE:

Ask children to use the ideas from the Explore section to help them find out more about water's importance for the health of animals, including humans, and also how water is transported in humans and how this is different in some other animals. Explain to children that they need to prepare a one-minute presentation to share **three** things they have learned, **two** questions they still have, and **one** thing that they will change in their lives after this lesson. The challenges are differentiated by the level of support provided and the amount of information children need to collect and analyse.

Challenge 1: Children work in pairs to find out about how water is transported and used in humans and in an animal that lives in a desert

Ask: *How is water transported in humans? How is water taken in? What do humans use it for? How do human bodies get rid of water they don't need? How is this different to an animal that lives in a desert?*

Ask them to include at least one use of water in humans in their presentations.

Key information:

Animals that live in deserts are specially adapted to conserve water. The camel saves water by not sweating and by allowing its body temperature to rise. The kangaroo rat gets all the water it needs from its food and does not need to drink. The water is released once the food is broken down within each cell. Desert predators, such as the fennec fox and the jackal, get the water they require from the bodies of the animals they kill and eat.

Key information:

Seawater is too salty for humans and most land animals, but animals that live near or in salt water have adapted so that they can pump out the extra salt while keeping their salt levels in balance.

Challenge 2: Children work independently to find out about how water is transported and used in humans and in an animal that lives in a salt water environment

Ask: *How is water transported in humans? How is water taken in? What do humans use it for? How do human bodies get rid of water they don't need? How is this different to an animal that lives in salt water?*

Ask them to include at least one use of water in humans in their presentations.

Challenge 3: Children work independently to find out about how water is transported and used in humans, and identify an animal where the transportation of water is different

Ask: *How is water transported in humans? How is water taken in? What do humans use it for? How do human bodies get rid of water they don't need? Can you identify and find out about an animal that has adapted to live in an environment where the water supply is different to ours?*

Ask them to include at least one use of water in humans in their presentations.

REFLECT AND REVIEW:

Use slide 7 of Slideshow 1 to prompt and time the 3,2,1 presentations from each class. Collect three lists on the whiteboard: How many different uses of water have they found out about? Which animals have they found out about? What are they going to change in their lives?

EVIDENCE OF LEARNING:

Listen to children as they present their 3,2,1 presentation. Do children explain how water is used and transported in humans? Do they explain how water is transported and used in another animal? Do they identify the importance of water for human health?

CROSS-CURRICULAR OPPORTUNITIES:

There are links to keeping healthy in PSHE.

BODY PUMP

LESSON 7: WHAT DOES THE ROAD AROUND OUR BODY LOOK LIKE?

LESSON SUMMARY:

In this lesson children reflect on what sources of evidence they have used to learn about the human circulatory system and demonstrate their understanding in a card sort and by making a game. It is a good opportunity for self and teacher assessment of understanding of the human circulatory system.

Key vocabulary:

circulatory system, heart, lungs, rest of body, blood, veins, arteries, vessels, capillaries, valves, oxygen, nutrients, water, health, transport

Resources:

Scissors, split pins, modelling clay, access to all materials produced in Lessons 1–6, plus secondary sources of information that children have used previously in the module

Health and safety:

Remind children of the importance of using a ball of clay or tack when piercing a hole in the wheels to insert the split pin.

Key information:

If children are unfamiliar with a dial like this, show them that the explanation goes in the box directly opposite the label, not the one under the heading, as the arrow lines up with the word but the aperture is opposite.

National curriculum links:

Identify and name the main parts of the human circulatory system and explain the functions of the heart, blood vessels and blood; to describe the ways in which nutrients and water are transported within animals, including humans

Learning intention:

To create a game to demonstrate knowledge of the human circulatory system

Working scientifically links:

Identifying evidence that has been used to support and refute ideas or arguments

Success criteria:

- I can label a diagram of the circulatory system.
- I can explain the function of the different parts of the system.
- I can self-assess my reporting skills.

EXPLORE:

Begin by watching the schoolhouse rock circulation song together (http://www.youtube.com/watch?v=1_22fZdI_vA). Ask children if there was anything there that they did not already know.

Play a game of True/False using Resource sheet 1. Children should work in small groups of no more than four. Check the answers together using the table provided. Ensure there are no continuing misconceptions before moving on to the challenge activities.

ENQUIRE:

Explain to children that they are each going to make a circulatory system body wheel to remind them of all they have learned about how gases, nutrients and water are transported throughout the body. The challenges are differentiated by the level of support given and the complexity of information required.

Challenge 1: Children make a circulatory system body wheel

Use Resource sheet 2 as a template. Children should colour the blood vessels in red and blue according to scientific convention, draw lines from the names to the correct parts and write short explanations of what each part does. They can add and label any more parts they know.

Challenge 2: Children make a circulatory system body wheel, adding extra labels and two more explanations

Use Resource sheet 3 as a template. Children should colour the blood vessels in red and blue according to scientific convention, label the parts and write short explanations of what the part does, adding two more of their own. They can also add and label any more parts they know.

Challenge 3: Children make a circulatory system body wheel, adding extra labels and four more explanations

Use Resource sheet 4 as a template. Children should colour the blood vessels in red and blue according to scientific convention, label the parts and write short explanations of what the part does, adding four more of their own. They can also add and label any more parts they know.

REFLECT AND REVIEW:

Use Interactive 1 (this is an interactive version of the True/false activity from earlier in the lesson) and ask children to think about how their wheels show some of the things that they have learned about the human circulatory system in this module.

Now ask children to reflect on what they have achieved in this module with reference to the following working scientifically statements. How do they feel this module has developed their skills? Encourage them to recognise which skills they need more practice with and to identify other subject areas where they might be able to develop the skill, such as English. Can they:

• report findings from enquiries including explanations of results?

• report findings in written and oral forms such as displays and presentations?

• identify scientific evidence that has been used to support or refute ideas or arguments?

EVIDENCE OF LEARNING:

Observe children closely during the lesson, focusing on the confidence with which they use scientific language to explain the function of the different parts of the circulatory system. Do they know what the main parts are and what they do? Do they refer to the evidence they have used during the module to develop their knowledge? Do they self-assess realistically the strengths and weaknesses in their reporting and presentation skills?

BODY HEALTH

INTRODUCTION

In this module children learn about how to keep their bodies healthy and how their bodies might be damaged. The focus is on lifestyle choices that humans make, including diet, exercise and drug use, and how these are informed by scientific evidence.

Children will build on their learning from Year 3 about the types of food that humans and other animals need in order to stay alive. They will develop a deeper understanding of what constitutes a healthy diet, through exploring food groups and how the body uses them. They investigate food packaging to find out what snacks and drinks contain, and use this information to inform their own choices of drinks and snacks. The children also investigate how the results of scientific enquiries have influenced what we eat.

In addition, the module draws on children's learning in Year 3 about the functions of the skeleton and muscles. They explore the effects of exercise on the body and develop their understanding of the circulatory and respiratory systems as they investigate the effects of exercise on the pulse and its recovery rate. They then find out about the training regimes of athletes and learn about about special diets and training programmes.

Children will have the opportunity to find out about how drugs help us as well as cause us harm. There is a clear link with work in PSHCE, and it is recommended that any school project on health is planned in conjunction with science lessons.

Care should be taken to treat this topic with sensitivity throughout. Knowing any dietary and health issues of pupils is vital to ensure that any issues can be addressed carefully. Awareness of cultural, religious or health-based limitations on diet is essential and these should be treated sensitively to ensure that children understand one another's diets without judging them. This should be linked with PSHCE (awareness and understanding of cultural differences).

There are links to several lessons in Module 2, Body Pump, which should be taught first.

Key vocabulary:

alcohol, asthma, athlete, balanced diet, beats per minute (bpm), benefits, breathing, caffeine, calories, cancer, carbohydrates (including sugars), cheating, cigarettes, clinical trial, consequences, dairy, diet, doping, drugs, eatwell plate, energy, exercise, fat, fibre, heart, heart rate, intensity, illegal, impact, James Lind, legal, lifestyle, long-term effect, lungs, medicine, mental benefits, mineral, motivation, norm, nutrition, oxygen, passive smoking, peer pressure, performance enhancing, persuade, physical benefits, protein, pulse rate, RDA (recommended daily allowance), recovery rate, resting rate, rickets, roughage, saturated fat, scurvy, short-term effect, smoking, sodium, solvents, steroids, tobacco, training, unsaturated fat, vitamin

FACT FILE:

The health of humans can be adversely affected by the following:

- A poor diet: A healthy diet is one that helps to maintain or improve general health, providing the body with essential nutrition, including water, protein, essential fatty acids, vitamins, minerals and adequate energy (expressed in calories). A healthy diet can include plant- and animal-based foods. Where there are no pre-existing health problems, a properly balanced diet (in addition to exercise) is also thought to be important for lowering health risks such as obesity, heart disease, type 2 diabetes, hypertension and cancer.

  This must be handled sensitively, as some children or family members may suffer from these conditions, which are not caused only by diet.

- Exposure to disease-causing micro-organisms: Micro-organisms can be transmitted to and between humans in several ways, including eating and drinking contaminated food and water, through coughs and sneezes, by direct contact and by disease-carrying organisms such as mosquitoes and fleas.

- Exposure to harmful substances: These include tobacco, which has been directly linked to breathing disorders, blocked arteries, heart disease, lung and other cancers and nerve damage, and alcohol drug and solvent abuse, which have been directly linked to impaired performance, personality change and major organ damage.

- Lack of exercise, rest and sleep: Regular exercise makes humans stronger and more efficient, and a lack of regular exercise can lead to joint and muscle problems, clogged arteries, high blood pressure and heart disease. Humans need to rest and to sleep so that the body can repair and recharge itself. Insufficient sleep can lead to stress, anxiety and impaired performance.
- Stress: Stress can be caused by a wide range of physical, emotional and environmental factors, and lead to a range of physical and physiological symptoms.

Common misconceptions:

Children generally attribute good health to what they eat and drink, and identify individual foods as healthy, rather than recognising the need for a balanced diet or eating in moderation. Children often see exercise and rest as just adult pursuits.

Alternative ideas that children may hold about diet include:

- Overweight people are unhealthy and slim people are healthy. Children may not recognise that there is a range of healthy body weights for people of any height.
- All fatty foods are bad for you. Children may not understand that the body needs some fat, and that it is a diet that is too high in fatty foods that can lead to people becoming overweight or obese.
- Childhood obesity is the parents' fault. It is unhelpful to put the blame on just the parent if a child is overweight. Families need to work together to overcome childhood obesity before it becomes adult obesity. The adult who buys the food is responsible for making choices that are as healthy as possible, and this includes suitable portion sizes. Helping children to understand the reasons for these choices will help them to make healthy choices when they eat at a friend's house, choose from the school lunch menu or eat at a restaurant.
- Childhood obesity is caused by fast food. No single food type causes obesity, but eating high-calorie food in large quantities over time is likely to lead to the child becoming overweight. Fast food, restaurant food and shop-bought food can have high calorie levels. Eating them without an awareness of what they contain in relation to what the body needs can cause problems with weight. Fish and chips don't make you fat, its the decision to eat too much too often, especially when combined with a sedentary lifestyle, that means the body is taking in too many calories, the excess of which will be stored as fat.
- Childhood obesity is caused by too much TV and video games. Watching TV and playing computer games don't make children fat, but prolonged periods of inactivity can lead to weight gain if the body isn't doing enough exercise to use all the energy store it takes in as food. It is important that adults limit TV time and encourage more active time outdoors. Doing this as a family can be helpful to the health of the entire family.

Big Cat book links

Extreme Sports Adrian Bradbury 978-0-00-733632-6 Band 14 Ruby	Find out about the dangers people face in extreme sports
My Olympic Story Kwame N. Acheampong 978-0-00-733636-4 Band 15 Emerald	Follow Kwame's amazing life.
Ade Adepitan: A Paralypain's Story Ade Adepitan 978-0-00-746548-4 Band 16 Sapphire	Paralmpic medal winner and successful TV presenter, Ade Adepitan has led a remarkable life.
My Journey Across the Indian Ocean James Adair 978-0-00-7465521 Band 17 Diamond	Sharks, huge waves and months at sea – these are just a few of the things that James and Ben had to battle against.
Becoming an Olympic Gymnast Beth Tweedle 978-0-00-742837-3 Band 18 Pearl	What's it like to be one of Britain's greatest gymnasts?
Swimming the Dream Ellie Symonds 978-0-00-742838-0 Band 18 Pearl	Find out about Paralmpian Ellie's story of professional swimming and living with Achondroplasia Dwarfism.
The Tour de France Sean Callery 978-0-00-742839-7 Band 18 Pearl	Find out how the Tour began, what stages are involved and what prizes can be won.

BODY HEALTH

 ## LESSON 1: WHAT DOES BEING HEALTHY MEAN?

LESSON SUMMARY:

In this lesson children revise their learning from Year 3 about how humans obtain nutrition from the different types of food they eat. By the end of the lesson they are able to describe how both diet and exercise contribute to a healthy lifestyle.

Key vocabulary:

diet, food, exercise, healthy lifestyle, impact, nutrients, water, oxygen, carbohydrates, fats, proteins, minerals, essential, healthy, vitamins, regular, calories, balanced

Resources:

Large sheets of flip chart paper, sheets of A3 paper, pens, glue, scissors, plain A4 paper, access to secondary sources, including the internet and healthy education pamphlets and posters, for all levels of challenge

National curriculum links:

Recognise the impact of diet, exercise, drugs and lifestyle on the way bodies function

Learning intention:

To describe the impact of diet and exercise on human health

Scientific enquiry type:

Finding things out using a wide range of secondary information

Working scientifically links:

Reporting and presenting findings from enquiries, including conclusions, causal relationships and explanations of and degree of trust in results, in oral and written forms such as displays and other presentations

Success criteria:

- I can explore the link between diet, exercise and a healthy lifestyle.
- I can present my findings as a concept map or poster.

EXPLORE:

Remind children that in Year 3 they learned about how their bodies need different types of food to stay alive. Explain to them that in this module they are going to investigate in more detail what humans can do to look after their bodies and be healthy.

Give pairs of children the pictures from Healthy or not? (Resource sheet 1). Ask children to sort the pictures into two groups according to whether they think the activity or behaviour is healthy or unhealthy. Make sure that children consider the activity or behaviour, not the person's appearance.

Ask each pair of children to compare their picture sorting with another pair.

Ask: *Do you disagree about any of the activities? Why? Does it make a difference if the person does something every day or just occasionally?*

Now ask them to sub-divide the healthy and unhealthy piles according to whether they are connected to diet or exercise.

Key information:

A healthy diet is one that helps maintain or improve general health. A healthy diet combined with exercise is thought to reduce the risks of health problems. A healthy diet can be met through a balance of plant-based and animal-based foods.

ENQUIRE:

Explain to children that their challenge is to produce a concept map or poster to show others what they know or can find out about a healthy lifestyle. Use your observations of children's responses to the Explore part of the lesson to help you group children for either Challenge 1, 2 or 3. Challenges 1 and 2 ask children to create a concept map with differing levels of support and Challenge 3 involves the creation of a poster, offering children an open-ended opportunity to share what they already know beyond the vocabulary provided for Challenges 1 and 2. All groups work to develop their understanding of a healthy diet. Careful observation of all groups and the work they produce will allow you to identify any misconceptions children may have about diet, exercise and obesity.

Challenge 1: With support, children create a concept map about leading a healthy lifestyle

Work with this group of children (if more than six pupils, split them into smaller groups) to develop a concept map using the Concept vocabulary (Resource sheet 2). Ask the children to make links between the words, such as linking the three cards 'diet', 'exercise' and 'healthy' to make the sentence: A balanced diet and moderate exercise help you to lead a healthy lifestyle. It may be useful to scribe the links for the children if this allows more thinking time for them. During this activity, help the children to unpick any misconceptions they have, where possible addressing them through peer support, a secondary source, a model or a diagram. As the children develop their map, introduce one statement at a time from Rumour mill (Resource sheet 3). Ask the children if they agree or disagree with the statement. Can they use secondary sources to find evidence to support or refute it?

Challenge 2: Children create a concept map to explain what they know about leading a healthy lifestyle

The children work independently to stick the word cards from Concept vocabulary (Resource sheet 2) onto a large sheet of paper in close proximity to each other, draw lines, add annotations and any new words they wish to show what they understand about being healthy, and the links between diet, exercise and other variables. As the children work on their concept map, offer them regular opportunities to select a rumour from Rumour mill (Resource sheet 3). They should search for evidence to support or refute the rumours and add information to their concept maps to demonstrate these as facts. For example: Childhood obesity is the parents' fault. The children's research should uncover that parents have a responsibility to buy food that provides a balanced diet and of suitable portion size but that blame is not overly helpful.

Challenge 3: Children work independently to develop a poster to show what leading a healthy lifestyle means

They should include the importance of diet, exercise and other variables that they believe are important. You may wish to give them a copy of Concept vocabulary (Resource sheet 2) for prompts, but encourage them to present their understanding in their own way using scientific vocabulary. They should explain carefully any diagrams or images they use in their poster. In addition, provide them with a copy of Rumour mill (Resource sheet 3), and access to secondary sources. They should test these statements when planning their posters in order to add information that supports or refutes the statements.

REFLECT AND REVIEW:

As a class, present one, or more if time allows, rumours from Rumour mill (Resource sheet 3), and ask children to share evidence with a partner to support or refute the rumour(s) based on what they have found out today. Encourage children to talk to children from other challenge groups as well as their own group. Posters and concept maps could be displayed in the classroom as part of this module. Together as a class, ask children to use the concept maps and posters to define the indicators of being healthy.

EVIDENCE OF LEARNING:

Observe and listen to children as they sort the diagrams. Review the concept maps and posters. Does their work show clear understanding of the link between diet, lifestyle and health? For example, do children link fats and heart disease, sugars and obesity, and lack of exercise with being unhealthy? Can they define the indicators of a healthy lifestyle, such as a balanced diet, coupled with moderate exercise? Are children able to justify their opinions supported by their research? For example, do they summarise information from secondary sources rather than copy it?

BODY HEALTH

 ## LESSON 2: HOW IS FOOD DIVIDED INTO DIFFERENT GROUPS?

LESSON SUMMARY:

In this lesson children examine food packaging labels to identify the food groups that different types of food contain, using their existing knowledge of the four main food groups from Year 3. They evaluate guidance that is given to people looking to plan a healthy diet. Ensure children are clear about the scientific use of the word 'diet' in this context. We are not referring to the choices people make when trying to lose weight, but to the usual food and drink that someone consumes.

Preparation required:

This lesson requires food packaging labels. These should be from a range of food types and sources. You could ask your class to bring in empty clean packaging in advance of this lesson or, alternatively, use some of your own. You will need at least the same number of food items as children, ideally double that to allow them to work at their own pace rather than waiting for another child. You may also wish to keep some of the packaging, as more is required in Lesson 3. A good secondary source of information for all levels of challenge is http://www.nhs.uk/Livewell/Goodfood/Pages/Healthyeating.aspx.

National curriculum links:

Recognise the impact of diet, exercise, drugs and lifestyle on the way bodies function

Learning intention:

To evaluate healthy eating guidance

Scientific enquiry type:

Grouping and classifying things

Working scientifically links:

Identifying scientific evidence that has been used to support or refute ideas or arguments

Success criteria:

- I can use packaging information to sort foods into the different food groups.
- I can use health guidelines to plan a healthy menu.
- I can say how guidance about food content is helpful.

EXPLORE:

Show children the Eatwell plate (Resource sheet 1).

Ask: *What do you think this is for?*

Either as a class using the IWB, or individually or in pairs using a laptop, children play Sort the food types (Interactive 1). This is designed to be simple and a quick check of knowledge from Year 3. As children move each type of food to the various boxes on screen, ask them if they know which food group contains which food and why we need to eat foods from each group.

You may need to remind them that the different food **groups** provide the nutrients that humans need to live and grow. The body gets energy from food and oxygen. Establish that children understand that a calorie is a unit of energy and that without sufficient calories our bodies would not function. The four main food **groups** are:

proteins that help our bodies grow and repair themselves

carbohydrates that give us energy

vitamins and *minerals* that are good for our skin, bones, teeth and blood

fats that provide energy and help in building our bodies.

This should be familiar to children from Year 3 when they learned about how the different **types** of food – fruit and vegetables; starchy foods including breads and pasta; meat, fish, eggs and beans; dairy products; and sugars – are sources of the different food **groups**.

Key vocabulary:

carbohydrates (also referred to as starchy foods), proteins – including meat, fish, eggs and beans, fats, sugars, fibre, calories, dairy, RDA (recommended daily allowance), saturated fat, unsaturated fat, salt/sodium, eatwell plate, vitamins, minerals, roughage

Resources:

Range of food packaging either sealed or clean and empty, mini whiteboards, small paper plate (one per child)

Key information:

Healthy eating guidance often uses the term 'food groups' as shorthand for the different **types** of food that provide the four main food groups.

ENQUIRE:

Explain to children that a great deal of money is spent helping people plan healthy diets. Leaflets and posters are produced, adverts are made and food packaging has large amounts of information. Their challenge today is to evaluate how useful this information is. Discuss the term 'recommended daily allowance'.

Challenge 1: Children use food packaging to identify which types of food contain the food groups needed for a healthy diet

The children work in groups of four. Each child chooses a food group from vitamins and minerals, carbohydrates and proteins and fats, and looks at food packaging to identify food that provides a high proportion of this food group. They then use the Eatwell plate (Resource sheet 1) to match the food they have identified to its food type. Where does it go on the plate? They share their findings with the others in their group to work out a day's balanced diet.

Ask: *How does food packaging help you plan a healthy diet? How does the food plate help? Are some food groups easier to find than others? Which food type have you eaten most of already today? What food group was that? What food groups should you eat more or less of?*

Challenge 2: Children use food packaging and other sources of information to design a packed lunch that contains the balance of food groups needed for a healthy diet

The children work in pairs. Using the Eatwell plate (Resource sheet 1) as their guide they plan a packed lunch which contains the right balance of the different food types. They then examine the information on the packaging of the food they have chosen and use other sources of information to work out how much of each of the different food groups their lunch contains.

Ask: *Is anything missing? How does the food plate help you to plan a healthy diet? Was the food packaging useful? Why do you think health leaflets focus on helping people to balance food types, not food groups?*

Challenge 3: Children use food packaging to identify how much fat and sugar is in foods

The children work in pairs and use the Eatwell plate (Resource sheet 1) to identify the recommended daily allowance of fats and sugars.

Ask: *Can you find out why is it smaller than the recommendation for other food types?*

The children then look at the information on food packaging to find out how much fat and/or sugar is contained in foods that they might think belong in other food types, such as ready meals, soup and breakfast cereal.

Ask: *Which foods are high in fats and sugar? Have you found anything that surprised you? Are health professionals right to talk about the 'dangers' of hidden fats and sugars? What evidence do you have?*

REFLECT AND REVIEW:

Give each child a small paper plate and ask them to draw a line down the middle. On one side ask them to write why information on food packaging is a good thing. On the other side they should write why the eatwell plate is a good idea.

EVIDENCE OF LEARNING:

Observe and listen to children as they talk about the different food groups. Look at their presentations. Do children sort food into the correct types? Do children know what the main food groups are? Can they describe a balanced diet in terms of food types and/or food groups? Do they use food labels and other sources of information to identify the types of food that are good sources of each of the different food groups? Can they evaluate the value of food labelling and healthy diet guidance such as the eatwell plate?

BODY HEALTH

LESSON 3: WHAT MAKES A HEALTHY SNACK OR DRINK?

LESSON SUMMARY:

In this lesson children build on their learning from Lesson 2 to examine the nutritional content of certain snacks and drinks to decide whether they would contribute to a balanced, healthy diet. They examine different food packaging and look at how carefully worded food packaging can sometimes be misleading to the purchaser. They apply their knowledge to snack and drink packaging design. It is assumed that children are familiar with grams and millilitres, and methods of recording these measurements. Some children also need to be familiar with working with one or two decimal places (Challenges 2 and 3). By the end of this lesson children will have identified a list of healthy drinks and snacks, and reflected on their own diets as a result of this.

Preparation:

You will need a large collection of snack food and drink packaging for children to use in this lesson. Ask colleagues or children to help with this ahead of this lesson. You should aim for approximately three pieces of snack or drink packaging per child. Try to ensure there is a balance of food and drink types such as cola and sugar-free cola. Food labels (Resource sheet 1) is provided for food and drink packaging that may not come with nutritional packaging, such as individual pieces of fruit.

Key vocabulary:

snack, drink, balanced diet, nutrition, energy, protein, carbohydrates, of which sugars, fats, of which saturates, fibre, sodium

Resources:

Sticky notes (five per child), wide range of snack food packaging with nutritional value information accessible

Health and safety:

The Reflect and review part of the lesson includes a 'tweet' as a way of sharing what children have learned. Ensure you check your school's policy on using Twitter. It is intended to be a pretend tweet, but if your school uses Twitter to share learning, a real tweet could be used.

National curriculum links:

Recognise the impact of diet, exercise, drugs and lifestyle on the way bodies function

Learning intention:

To identify criteria to judge whether a drink or snack is healthy

Scientific enquiry type:

Finding things out using a wide range of secondary sources of information

Working scientifically links:

Recording data using a table and reporting and presenting findings from enquires, including causal conclusions, relationships and explanations of and degree of trust in results, in oral and written forms such as displays and other presentations

Success criteria:

- I can use food packaging to find out what is in a food or drink product.
- I can compare this with what I know about a balanced diet.
- I can make decisions about which snacks I eat based on what I know.
- I can interpret information I have collated in a table and link it to my own diet.

EXPLORE:

Ask children to divide into pairs.

Ask: *What do you understand by the word 'snack'?*

Give them 1 minute to discuss. Collect responses from the class and agree a definition that includes the idea of it being a food or drink that is consumed beyond their normal three meals a day.

In order to collect data about the foods which the class snack on, give each child five sticky notes. Ask them to write down five typical snacks they eat or drink in a normal week, putting one on each sticky note. Create a large collection space in the classroom for children to bring their notes together and classify them. They may wish to use hoops on the floor, a large whiteboard or sheets of flipchart paper. They should then collectively decide names for each group they have created. Note: with larger classes, you may wish to do this in smaller groups rather than as a whole class.

ENQUIRE:

Explain to children that their challenge is to look in more detail at the food labels on the snack packaging and compare them to decide if they would make a healthy snack. Remind them of the work they did in Lesson 2 about the different food groups and what constitutes a balanced diet. In order to benchmark whether a snack is healthy or not, ask them to look carefully at the food packaging information, in particular the recommended daily allowance (RDA), which is usually expressed as a percentage. Discuss how, in the main, the lower the percentage, the more suitable the product is as a healthy snack.

Challenge 1: Children investigate the contents of a snack product and record the nutritional information to 1dp with support

Children find out more about what is in a snack, identifying foods that make good snacks and those that should be eaten less often. Ask them to select a snack product, look carefully at the food label and use the provided writing frame What's in a snack?: Challenge 1 (Resource sheet 2) to record their research. The children may have difficulty with decimal place value and may find it challenging to compare large amounts of data, so the resource sheet asks them to look at a limited range, recording only whole numbers. For any value below 1.0 ml or g, they should record a 0. Some support may be needed for this group to remind them of the work they did in Lesson 2 when they looked at food packaging in general, such as the difference between 'saturated fat' and 'unsaturated fat' or the difference between 'carbohydrates' and 'carbohydrates of which sugars'. Allow them to record data for several snacks.

Ask: *Looking at your table, which food would make a healthy snack? How do you know this? Which food is not so healthy? How do you know? How will this affect choices you make about snack food?*

Challenge 2: Children investigate five food and drink snack products and record their nutritional information to 2dp

It is assumed that the children doing this challenge are able to read and interpret information quickly. In particular, they need to look at fat, saturated fat and sugar to help them make decisions about the health value of the snack. They should also be comfortable comparing numbers to one decimal place. Working independently, ask the children to investigate five different food or drink snack products. They should record nutritional information per 100g and per portion size of the snack using the writing frame What's in a snack?: Challenges 2 and 3 (Resource sheet 3).

Ask: *Can you compare the nutritional information gathered and decide which snack is the healthiest? Can you rank them, with 5 being the healthiest and 1 the least healthy?*

Challenge 3: Children investigate and record nutritional information on five snacks and five drinks

It is assumed that children doing this challenge are confident working quickly with numbers to one or two decimal places and are familiar with all the main food groups. Ask them to work quickly to record information for five different snacks and five different drinks based on the food labels. They should use two copies of the writing frame What's in a snack? (Resource sheet 2) – one for snacks and one for drinks.

Ask: *Can you reflect on the data gathered and make a summative decision about what makes the most healthy snack and drink? Can you reflect on the foods you snack on and share any changes to your snack habits you plan to make?*

REFLECT AND REVIEW:

From their research, ask children to offer ideas for a class list of healthy snacks. Ask them to spend time in pairs looking at one piece of packaging. Ask them to consider any information that could be misleading to the consumer. For example: the words 'low fat' on a yoghurt may lead the consumer to assume it is a healthy snack. A more detailed look at the ingredients and the nutritional information may reveal a high number of calories per portion. Discuss the serving sizes used for the nutritional information. Packs tend to have the nutritional levels in one pack and/or 'per 100 g'. If a child eats two bags of crisps, what are the actual levels they are consuming? How does this compare with the RDA?

Ask children to 'tweet' what they have learned about snacking using 140 characters or fewer. If children are unfamiliar with this type of writing, you may need to allow them extra time to experiment. An example might include 'What makes a healthy snack? Did u know cheddars r a healthy snack? Low in calories & fat.' This could be done on a sticky note and added to the class display on an image of a crisp packet or chocolate bar. It may also be stuck in children's books as evidence of their learning.

EVIDENCE OF LEARNING:

Watch children as they sort their snacks in the Explore session and as they investigate the nutritional values on the packaging. Listen to children's reasoning as they choose which snacks are the healthiest. Can children interpret the nutritional values on food packaging correctly? Can they describe what makes a healthy snack? For example, raisins are not a particularly healthy snack as they are very high in sugar, while small cheddar cheese crackers are a healthy snack as they contain very few calories and very little saturated fat.

BODY HEALTH

LESSON 4: HOW HAVE DIETS CHANGED?

LESSON SUMMARY:

In this lesson children look at how ideas have been tested scientifically to identify cause and effect and how the results have impacted our diet. They investigate historical cases of diet affecting health, including scurvy and the work of scientist James Lind. By the end of this lesson they have considered how scientific ideas are developed and how scientists gather evidence to support this.

Key vocabulary:

scurvy, James Lind, rickets, history, health problems, poor diet, scientific idea, scientist, clinical trial, test

Resources:

Access to secondary sources for all levels of challenge

National curriculum links:

Recognise the impact of diet, exercise, drugs and lifestyle on the way bodies function

Learning intention:

To use secondary sources to investigate how scientific ideas were developed in the past

Scientific enquiry type:

Finding things out using a wide range of secondary sources of information

Working scientifically links:

Identifying scientific evidence that has been used to support or refute ideas or arguments

Success criteria:

- I can describe how diet can affect health.
- I can talk about at least one way scientists gathered evidence to test scientific ideas.
- I can use secondary sources to find information to support or refute my ideas.
- I can present my findings to others in a presentation.

EXPLORE:

Hand out one old wives' tale card to each pair of children from the selection provided on Old wives' tales (Resource sheet 1). In pairs, ask children to spend time talking about their old wives' tale.

Ask: *What message is it giving us about that particular food? Is it positive or negative? Do you know if there is any truth in it? What evidence do you have to explain your answer?*

Explain to them that there have long been ideas that link food to health. These are based on observed evidence and patterns identified. Most are based on a modicum of truth. For example, carrots really do give your eyes a boost because they contain beta-carotene, which the body is able to convert into vitamin A, an essential vitamin for healthy vision. Other vegetables such as sweet potato and apricots also contain high levels of beta-carotene.

ENQUIRE:

Explain to children that they are going to find out about times when we knew less about healthy eating and many people died because of diet-related illnesses. Explain to children that they are going to learn how scientists set about trying to work out what causes particular illnesses and how to avoid them in the first place or cure them if not. Show children the video about James Lind (Video 1). It includes a timeline and a brief outline of how he used the first ever clinical trial aboard a naval ship in the 1700s to identify the role citrus fruits had in aiding recovery from scurvy, thus almost eradicating it from the navy.

Explain to children that they are going to research the problems caused by poor diet in the past. Challenge 1 looks in more detail at the work of James Lind, Challenge 2 uses drama to tell the story of the first clinical trial and Challenge 3 investigates other diet-related illnesses that have been largely eradicated from the UK because of the work of scientists.

Useful starting points for pupils' research include: http://www.bbc.co.uk/history/historic_figures/lind_james.shtml

http://www.jameslindlibrary.org/illustrating/articles/who-was-james-lind-and-what-exactly-did-he-achieve

Challenge 1: Children record their findings on the work of James Lind on a writing frame

The children find out about the work of James Lind in more detail. Explain to them that they can watch the video again if they wish and that they have access to secondary sources such as books or the internet to carry out research. Give them each a copy of What did James Lind do? (Resource sheet 2) to record their findings. This will focus their research and help them think about the clinical trial in more detail as well as what the scientist went on to do after this piece of research.

Challenge 2: Children work in groups to research and present a four-part freeze-frame drama sequence

The children work together in small groups of three or four. Using what they know already from watching the video, and using secondary sources to undertake further research, explain to them that they are to present a four-part freeze-frame drama sequence.

Ask: *What happened first? What might that look like as a freeze-frame? What happened next? How do you know this? Can you summarise the main events of James Lind's clinical trial in four freeze-frame scenes?*

Give them the Scurvy freeze-frame plan (Resource sheet 3) to plan their freeze-frames and make notes when researching. They should present their work to other groups carrying out Challenge 2.

Challenge 3: Children carry out research and choose how to present their findings

This challenge relies on good reading, writing and presentation skills. The children research other historical dietary issues such as rickets as well as scurvy and look at the likelihood of them returning to our country again. They should choose how they wish to present their findings, such as in a slideshow, a poster or a table comparing the different diseases. They should research at least two historical diet and health-related issues and should include:

- dates
- key features of the health problem
- misconceptions of how they were caused at the time
- important scientists or researchers who discovered the reasons for the condition
- how scientists went about testing their scientific idea in practice
- what we have learned that affects our knowledge of a healthy diet today
- any current research that indicates this problem is likely to reoccur in their country or elsewhere in the world.

Key information:

This part of the lesson could link well with any links the school has with 'real' scientists. It may offer opportunities to invite a scientist into school or use one of the many online video link-ups that are available specifically for schools, such as 'I'm a scientist, get me out of here' http://imascientist.org.uk/

REFLECT AND REVIEW:

Remind children that James Lind's work is still significant today because he was the first scientist to carry out a clinical trial to test an idea. In pairs, hand out the old wives' tale cards from Resource sheet 1 again.

Ask: *Can you think of how a scientist might go about investigating this idea in a clinical trial? What might they test? How? What would they hope to find out?*

Collect some shared responses.

Ask: *Can you relate this idea to any investigations you have done in school?*

You may need to prompt some ideas where this has happened in your school, such as a plant-growing investigation linked to changing one variable in the growing conditions.

EVIDENCE OF LEARNING:

Observe children while they are working on their Enquire task. Do they include key facts from secondary sources that support their presentation? Do they link the work of James Lind with changes in the longer-term diet of people? Can they describe how what you eat has an affect on your health? Do they present their findings in a way that is appropriate to their challenge? Do they make connections with the way they carry out different types of science enquiry in school and the work of present day and historical scientists?

BODY HEALTH

LESSON 5: HOW IS PULSE RATE AFFECTED BY EXERCISE?

LESSON SUMMARY:

In this lesson children explore the impact of exercise on the body. They learn that they can measure their pulse rate to find out how hard their heart is working. They measure their resting heart rate and collect data to investigate what happens when they exercise. They make links with their learning about the circulatory system from Module 2, Body Pump. As it may take some time for children to learn how to take their pulse accurately, a valuable skill in itself, it is recommended that this learning could be spread across two lessons. By the end of this lesson children are able to interpret a graph of class data to describe how heart rate is affected by exercise.

Preparation required:

Ask around in your class prior to this lesson to discover if any children are part of elite sports training outside of school. Regional or even national standards of training will have taught these children lots about the link between resting rate, exercise and recovery rate.

Key vocabulary:

heartbeat, pulse rate, beats per minute (bpm), resting rate, stopwatch, exercise, heart, norm, recovery rate

Key information:

The term 'recovery rate' is referred to in this lesson. It should be explained as the time taken for the pulse to return to near normal resting rate after any exercise that raises the pulse rate.

Resources:

Stopwatch (one per pair)

Health and safety:

Care should be taken to ensure this is a suitable activity for everyone in the class, based on your knowledge of their health and medical issues.

National curriculum links:

Recognise the impact of diet, exercise, drugs and lifestyle on the way bodies function

Learning intention:

To investigate variables that affect pulse rate

Scientific enquiry type:

Carrying out comparative and fair tests

Working scientifically links:

Taking measurements, using a range of scientific equipment, with increasing accuracy and precision, taking repeat readings where appropriate; reporting and presenting findings from enquires, including degree of trust in results

Success criteria:

- I can measure my pulse rate accurately.
- I can repeat measurements appropriately.
- I can interpret data in a table, identifying patterns in resting pulse rate.
- I can identify how exercise affects pulse rate.
- I can calculate my own recovery rate after exercise.

EXPLORE:

Show children Heartbeat and exercise (Slideshow 1) and ask them to discuss, in pairs, the statements made about a person's heartbeat. They should decide which response they agree with and why.

Ask: *What is a pulse rate? Why is it important to take your pulse rate?*

Ask children to discuss in pairs where to take a pulse. Explain to children that the pulse is taken on an artery and not a vein. Ask children to talk to a partner for a moment about why they think that might be. Ask children if they know the names of any arteries. Remind them of what they have learned about the heart from Module 2, Body Pump.

Show How can I take my pulse rate? (Video 1). Afterwards, clarify that the resting pulse rate for babies and newborns is usually 120–160 beats per minutes (bpm), for under-12s it is usually between 80–110 and for adults it is usually 60–100 bpm. It should be made clear to children that these are normal ranges but there are of course exceptions. A resting pulse rate of below 50 for an athlete is considered perfectly healthy. Ask children to find their own pulse. Some children will require help with this.

ENQUIRE:

By the end of this activity, each group will have recorded the resting pulse rate in bpm and identified patterns in the class relating to the norm for children their age. They then all explore what happens to their pulse when exercising and how long it takes for their pulse to return to the resting rate.

Challenge 1: Children measure and record their resting and post-exercise pulse rates, with adult support

The children work in pairs to record their resting pulse rate. They do this at least three times to check they are accurate. One partner records the time on a stopwatch (30 seconds) whilst the other takes their pulse. They should double the pulse rate for 30 seconds to get their bpm.

They then swap over. They add their results to a group table. Use the Pulse rate table of results: Challenge 1 (Resource sheet 1) to support this.

Now children do a short burst of high-energy activity, such as jumping jacks, running on the spot, or dancing for 2 minutes. They then take their pulse again, using the same method as above, and record the rate. They take and record their pulse rate every 5 minutes until it has neared the resting rate.

At the end of this activity, working as a group with adult support as required, children identify how many of their group have a bpm in each range.

Ask: *In the context of a scientific study would this sample size be useful?*

Explain to them that researchers seek to collect as much data as possible, as it is easier to observe patterns in a large sample size. In a classroom setting, however, this is not possible, so their sample size is limited and any generalisations made should be treated with caution. Finally, the group adds their data to the class table using the Pulse rate spreadsheet (Resource sheet 4) to input their data.

Challenge 2: Children work in pairs to measure and record their resting and post-exercise pulse rates

Ask the children to work in pairs to record their own resting pulse rate. As in Challenge 1, they take it in turns to take their pulse rate or time their partner. One partner times the other for 20 seconds, whilst the other takes the pulse before multiplying by 3 to get their bpm.

Ask: *Why do you need to measure your pulse rate more than once?*

The children first take their own pulse rate several times to identify their 'norm'. This may be recorded informally first until they are certain of their range of normal pulse rate. Each pair then records their pulse rate three times on the Pulse rate table of results: Challenge 2 (Resource sheet 2) before working out their average pulse rate or mean.

Now the children do a short burst of high-energy activity, such as jumping jacks, running on the spot, or dancing for 2 minutes. They then take their pulse again, using the same method as above, and record the rate. They take and record their pulse rate every 5 minutes until it has neared the resting rate.

They then add their results to the group table using Class pulse rate results (Resource sheet 4).

Challenge 3: Children work in pairs to measure and record their pulse rate and calculate the mean

Ask the children to work in pairs to record their own pulse rate. It is assumed children doing this challenge already understand the importance of repeating test results to check for accuracy, can calculate the mean and round numbers to the nearest whole number. They begin by working together to record their pulse rate for 1 minute each. They do this three times and calculate the mean. They then record their pulse rate for 10 seconds and multiply by 6, record for 20 seconds and multiply by 3, and record for 30 seconds and multiply by 2. If each of these measurements is taken three times and the mean calculated for each one, the children can identify if any of these shortcuts affect the accuracy of their measurements. You may wish to provide them with the Pulse rate table of results: Challenge 3 (Resource sheet 3) to support this.

Now the children do a short burst of high energy-activity, such as jumping jacks, running on the spot, or dancing for 2 minutes. They then take their pulse again, using the same method as above, and record the rate. They take and record their pulse rate every 5 minutes until it has neared the resting rate.

They then add whatever bpm they consider to be their most accurate to the group table, using Class pulse rate results (Resource sheet 4).

REFLECT AND REVIEW:

Each group should add their results to the template spreadsheet to create a whole class overview. Ask children to talk to their partner about what they notice. Encourage children to respond with observations including evidence from the data, such as *"I noticed that half of our class of 32 have a resting pulse rate of between 85 and 94 and that is normal for our age"*.

Ask: *What happened to your pulse rate when you increased your physical output? Why? What is happening to the heart? What did you notice when you took your pulse 5 minutes after the exercise?*

Ask a group to share their results and together as a class, draw the shape of the graph to show time (x) and pulse rate (y) before, during and after exercising. Establish that children understand that the quicker their pulse rate returns to a resting pulse, the greater the fitness indicator.

Ask: *What is your recovery rate?*

EVIDENCE OF LEARNING:

Watch children as they take their pulse. Can children record their data accurately? Do children recognise the importance of repeat readings? Do children make links between their pulse rate and the norm for their age? Can they identify their recovery rate after exercise?

BODY HEALTH

LESSON 6: WHAT ARE THE BENEFITS OF SPORTS AND EXERCISE?

Key vocabulary:

exercise, benefits, impact, healthy lifestyle, sport, heart rate, pulse rate, participation, persuade, training, motivation, physical benefits, mental benefits

Resources:

Access to secondary sources, including the internet, for all levels of challenge: sticky notes

LESSON SUMMARY:

In this lesson children survey the range of sports played by their classmates, consider the importance of exercise for a healthy lifestyle and develop ways to encourage more people to take up a new sport. By the end of the lesson children will have written an advert to encourage other Year 6 pupils to try a new sport or exercise.

National curriculum links:

Recognise the impact of diet, exercise, drugs and lifestyle on the ways bodies function

Learning intention:

To identify the impact exercise has on the way the body functions

Scientific enquiry type:

Finding things out using a wide range of secondary sources of information

Working scientifically links:

Reporting and presenting findings from enquiries, including conclusions, causal relationships and explanations of and degree of trust in results in oral and written forms such as displays and other presentations

Success criteria:

- I can explain the benefits of exercise on the human body.
- I can recognise variables that improve participation levels in sport.
- I can persuade others to try a new sport or exercise.

EXPLORE:

Ask: *What sports do you participate in regularly?*

Give each child some sticky notes and ask them to write each sport they regularly participate in on a separate sticky note. Gather these responses by arranging them as a graph where everyone can see it.

Ask: *What sport in our class has the most participants? Why do you think that might be? Are there any sports here that no one else has tried? Which is your favourite sport and why? What encouraged you to take up your favourite sport?*

Show the class Sporting quotes (Slideshow 1), which shows a series of quotes from the world of sports about motivation, engagement, fitness and determination. Ask children to pick one athlete and consider them.

Ask: *What motivates them to take part? Why do you think that is? What are the benefits of sport and exercise? How do you think people choose what sport to do?*

ENQUIRE:

Explain to children that they are going to link what they already know about healthy lifestyle, food, exercise, pulse rate and the circulatory system (from Module 2, Body Pump) and write an advert for a new sport, encouraging other Year 6 children to take part. Children doing Challenge 1 have a writing frame to support them, those doing Challenge 2 use some prompts and children completing Challenge 3 select their own presentation method and research a new extreme sport to engage with.

Challenge 1: Children write an advert for sport using a writing frame

The children compose an advert with support from a writing frame. Share the Try something new table (Resource sheet 1) and explain to the children that, independently, they are going to write an advert to encourage other Year 6 children to try a new sport. Ask them to select a sport or activity they already take part in or have done in the past so they have some initial knowledge of what is involved.

Ask: *What does the sport involve physically? How does that benefit the way our bodies function? How do you know this? How can you explain this to others? Can you think of how you could persuade your classmates to try it out?*

Ask them to use secondary sources to support their writing.

Challenge 2: Children create a poster advert for a new sport using prompts

The children use prompts to make a poster. Share Try something new (Resource sheet 2) with its prompts and explain to the children they are going to create a persuasive poster to encourage Year 6 classmates to participate in a new sport. Ask them to select a sport or exercise they may know a little about although not necessarily have tried themselves.

Ask: *What does the sport involve? Why would it be a good sport to take up?*

Encourage them to refer to all the prompts when writing their advert. Ask them to refer to secondary sources carefully to ensure they use facts to give reasons for taking part and opinions to persuade and encourage motivation.

Challenge 3: Children research a new sport and choose their own method of presentation for advertising

The children write a persuasive advert for a new or extreme sport of their choice.

Ask: *Have you included the mental and physical benefits of the sport? Have you provided enough evidence about the sport to persuade others to take part? Have you referred to secondary sources to support your research?*

REFLECT AND REVIEW:

As a whole class, ask children to find a partner who has written about a different sport or activity from them and to read what they have written to try and persuade them to try something new.

When sharing, ask them to give their partner marks out of five. One mark to be awarded for including each of the following:

- Explaining the benefits of exercise on the human body
- Including appropriate vocabulary such as pulse rate, heartbeat, calories and training
- Using tactics to persuade new participants to join in, such as a quote from a sportsperson or someone who regularly plays the sport
- Using facts about the sport or exercise from secondary sources
- Succeeding in making you want to give it a try!

EVIDENCE OF LEARNING:

Looking at children's adverts, use the same criteria children used as part of the Reflect and review session. Can they explain the benefits of exercise on the human body? Did they include appropriate vocabulary such as pulse rate, heartbeat, calories and training? Did they use appropriate tactics to persuade new participants, such as using a quote from a famous sportsperson or someone who regularly plays the sport? Did they use facts about the sport or exercise from secondary sources?

CROSS-CURRICULAR OPPORTUNITIES:

There are possible links with English in writing for an audience.

BODY HEALTH

 ## LESSON 7: HOW DO DRUGS AFFECT THE BODY OVER TIME?

LESSON SUMMARY:

In this lesson children explore the impact of drugs on the way the body functions. By the end of this lesson children have investigated the longer-term health effects on the body of taking drugs.

To maximise the impact of this lesson and Lesson 8, it is suggested that this module is taught at the same time as other drugs education work within the PSHE curriculum in order that deeper thinking and discussion around issues raised in this lesson can be addressed appropriately for children in your class.

National curriculum links:

Recognise the impact of diet, exercise, drugs and lifestyle on the way bodies function

Learning intention:

To identify and present the long-term effects on the body of drug use

Scientific enquiry type:

Finding things out using a wide range of secondary sources of information

Working scientifically links:

Presenting findings including causal relationships in oral and written forms

Success criteria:

- I can research the long-term effects of drug use.
- I can research the short-term effects of drug use.
- I can adapt information from secondary sources to help my presentation.
- I can share my findings with others in a visual way.

EXPLORE:

Explain to children that in this, and following lessons, they are going to learn about the longer-term effects of drugs and other lifestyle choices on the body. Establish that children understand what lifestyle means, that is, a mixture of elements including diet, exercise, drug usage, smoking, plus other variables such as stress, rest, happiness, fresh air and sleep. Explain that they are going to start with a quiz to see what everyone already knows.

Display Quick quiz – How do drugs affect the body? (Interactive 1) and ask children to record their answers independently. The answers are provided on screen after each question and on Quick quiz answers (Resource sheet 1).

In pairs, ask children to discuss the question: *How can medicines help us?* Possible answers may be: to help us get better; cure us; stop us from getting more unwell.

Ask: *Can you think of any dangers of taking them over a long period of time?*

Answers may include: dependence and over-reliance on drugs rather than lifestyle or diet change; the drugs may not be so effective over a long period of time.

ENQUIRE:

Remind children of the three drugs referred to in the quiz – alcohol, caffeine and solvents – and explain to them that they are going to investigate these and their effects on the human body in more detail. Explain to children that they are going to work independently to present findings of research into the longer-term effects of drugs on the human body. Children can choose which topic to undertake, combining their preferred subject matter and presentation style.

Challenge 1: Children describe long-term effects of drugs

Children use the How do drugs affect the body over time? writing frame (Resource sheet 2) to describe and present visually the longer-term effects of the drugs.

Key vocabulary:

drugs, medicine, illegal, legal, alcohol, caffeine, solvents, short-term effects, long-term effects, consequences, peer pressure

Resources:

Existing drugs resources your school may already have, access to secondary sources for all levels of challenge

Key information:

Drug education should always start at the level of the existing knowledge of the child. Observe carefully children's responses during the quiz in the Explore session and group your class appropriately based on their existing knowledge of drugs, and challenge them within that task as detailed.

Challenge 2: Children design a poster including information on drug legislation

Children design a poster highlighting the longer-term health effects of the drugs. The children include laws associated with the drug and advice for their peers.

Challenge 3: Children prepare a presentation on the effects of drugs

The children prepare a presentation of their choice, perhaps a slideshow or a comic strip, that includes a real-life case study of an adult or young person who has suffered the side effects of the drug being researched.

Topic 1:

Research the long-term health effects of drinking alcohol using secondary sources.

Short-term effects include: relaxation, headache, depression, loss of self-control, inability to speak clearly, feeling sick, taking stupid risks and losing balance.

Longer-term effects include: brain damage, liver damage, weight gain which can lead to diabetes, cancer of the mouth and throat, memory loss and possible stroke.

Topic 2:

Research the long-term health effects of drinking caffeine drinks using secondary sources.

Short-term effects include: stimulation of the heart, respiratory system and nerves, raising of blood pressure and relief of tiredness.

Longer-term effects include: raising the level of fatty acids in the blood, sleep deprivation, headaches, nervousness, agitation, shaky hands and palpitations.

Topic 3:

Research the long-term health effects of using solvents using secondary sources.

Short-term effects include: depression, giggling, dreaminess, dizziness, wooziness, hallucinations, nausea and vomiting.

Longer-term effects include: possible fatality, loss of appetite, frequent and persistent headaches, mood swings, 'chemical' smell on breath, spots around mouth and eyes, runny nose and damage to brain, liver and kidneys.

REFLECT AND REVIEW:

Ask children to write three long-term effects of using the drug they have researched on a sticky note. These could be added to the class display.

Ask: *At home, can you find out about the short- and long-term effects of smoking and the laws associated with buying and smoking cigarettes?*

Explain to them that they will need this information at the start of Lesson 8.

EVIDENCE OF LEARNING:

Observe children as they research the effects of the drugs. Do they include the short- and long-term effects on the human body of the drug they are researching? Can they differentiate between the two? Does their work visually communicate what they have found from their research?

Key information:

See http://www.cornwallhealthyschools.org/drugs-ed/drugs/drug-principles/ for useful information on principles for good drugs education.

BODY HEALTH

 ## LESSON 8: HOW DOES SMOKING AFFECT THE BODY?

LESSON SUMMARY:

In this lesson children investigate the risks posed to health by smoking. They explore the laws associated with smoking and the short- and long-term health risks associated with smoking. By the end of this lesson children are able to recognise how the influence of peer pressure can persuade people to do something dangerous or illegal, with smoking used as an example.

National curriculum links:

Recognise the impact of diet, exercise, drugs and lifestyle on the way bodies function

Learning intention:

To describe the long-term effects on the body of smoking

Scientific enquiry type:

Finding things out using a wide range of secondary sources of information

Working scientifically links:

Reporting and presenting findings from enquiries, including conclusions, causal relationships and explanations of and degree of trust in results, in oral and written forms such as displays and other presentations

Success criteria:

- I can identify the short- and long-term risks to my health of smoking.
- I can use my knowledge of the risks to my health to support my argument for not smoking.
- I can present my findings to others using scientific language.

Key information:

Ideally this and Lesson 7 should be taught as part of a wider drugs education program for children at your school and should be tailored accordingly.

Key vocabulary:

smoking, tobacco, cigarettes, lungs, cancer, breathing, asthma, passive smoking, peer pressure

Resources:

Two large PE hoops; access to secondary sources about smoking risks, should children wish to use them

Health and safety:

Be aware of children in your class whose parents or carers smoke. This lesson is not intended to frighten them but to give them a greater understanding of the risks and how they can be better prepared to make healthy choices as they grow up. In some rare circumstances, it may be known by the school that some children in Year 6 have already tried smoking. This should be handled sensitively and appropriately based on polices and guidance from your individual school.

EXPLORE:

Ask: *Can you tell someone one fact you found out about smoking in preparation for this lesson?*

Summarise this discussion with the following facts:

- Legally shops cannot sell cigarettes to people under 18 (as of 1 October 2007)
- Legally you cannot buy cigarettes until the age of 18 (as of 1 October 2007)
- Legally you cannot smoke in enclosed public spaces (as of 1 July 2007)
- It is illegal to display cigarettes in a large shop (as of 15 April 2012) and will be illegal in all shops from April 2015.

Ask: *Why do you think these most recent laws about point of sale advertising have been brought in?* Use the following information from www.ash.org.uk to add to suggestions the children identify.

Tobacco displays attract new young smokers and those who have given up or are trying to.

Tobacco display is a form of advertising. Tobacco advertising has been illegal in England since early 2003.

Public opinion since 2009 has shown strong public support for the ban of displays at the point of sale.

ENQUIRE:

Ask: *Can you think of any long-term or short-term health risks of smoking?*

Ask children to write each risk on a separate sticky note. Collect these either per table or as a whole class and sort them into short- and long-term risks to the health and wellbeing of the human body.

Short-term effects: feel less stressed and anxious, smelly clothes, hair and breath, stained fingers and teeth, coughing, increased heart rate and high blood pressure.

Longer-term effects: lung damage, lung cancer, heart damage, appear much older, cancer of the mouth, throat and lips, pregnancy complications, passive smoking can cause asthma and cancer.

Explain to children that their challenge today is to use their knowledge regarding the impact that smoking has on the human body to persuade others not to smoke. The challenges approach

Key information:

It is inevitable that some children will associate the cost of smoking with this task. Although this is indeed a reason not to smoke, remind them that for this lesson we are focusing on health risks to the human body. Cost implications could be investigated at another time.

this topic in different ways. Challenge 1 uses various scenarios to help children assess the risk to their health. Challenge 2 asks children to write a response letter as an agony aunt using scientific information. In Challenge 3 children role-play a phone call to support a family member.

Challenge 1: Children consider a risk associated with smoking and write a paragraph on it

The children differentiate between levels of risk associated with smoking. Give the children cut out cards from the Assess the risk cards (Resource sheet 1) and two large PE hoops, one labelled 'slightly risky' and the other labelled 'very risky'. In turns they take a card and place it in a circle.

Ask: *Why have you put that there?* For example: not washing – slightly risky as I might smell a bit and people may not want to work with me or stand next to me in the line for lunch. *Do the others in your group agree with your choice?* Finally offer them all a scenario card about an older friend offering them a cigarette. Give them time to discuss their views in pairs before writing, independently, a paragraph on what sort of risk they believe this to be and why. Encourage them to focus on the health risks, but to also think about the difficulty of saying no.

Challenge 2: Children explain, via letter, the risks associated with smoking and how to deal with peer pressure

The children consider the risks associated with smoking and suggest ways of dealing with pressure to do so. Explain to the children that they are going to become an agony aunt for a young person's website. Provide them each with a copy of Smoking: my story (Resource sheet 2). This is a letter to Cyber Aunt Sally from Alice, an 11 year old who has recently tried smoking. After discussion, ask them to write a response to her focusing on the health risks associated with smoking and suggesting ways she could deal with the peer pressure. Allow the children to discuss.

Ask: *What subtle ways can you think of that pressure is applied to smoke?* Answers may include celebrities smoking, parents and older siblings smoking, and a desire to show how mature they are. Ask them to remain focused on the scientific evidence when writing their letter.

Challenge 3: Children role-play a phone call to someone facing peer pressure to smoke

The children role-play scenarios which they create themselves or carry out using Phone call scenarios (Resource sheet 3). Ask the children to work in pairs to play the role of a supportive aunt or uncle on the phone and a child facing a smoking-related peer pressure scenario. If time allows, the children could create these scenarios themselves, as they are undoubtedly more realistic and relevant to their situation. If there is no time to create personalised scenarios, some have been provided in Phone call scenarios (Resource sheet 3). They may wish to spend time at this stage with access to secondary sources such as the internet or NHS smoking leaflets to gather information to help them respond with scientific evidence rather than family empathy. The children should role-play back-to-back to recreate a phone call where they cannot see the caller's facial expressions. Ask the children to reflect briefly in writing after each conversation.

Ask: *Did you use scientific evidence in your advice? How did that help you feel prepared to deal with the situation?*

Key information:

Knowing your own class, you should decide if children in this group would be more comfortable and honest with a partner of their choosing or if you should match up the pairs.

If they were the caller, ask them to write what they are going to do if they are faced with this scenario again. If they were the helpful relative, note what advice they gave. There is no expectation for each pair to use all the scenario cards, as the depth of conversation is more important than how many they have.

REFLECT AND REVIEW:

Pose the question from the lesson title: *How does smoking affect the human body?* Ask children to write down one effect on the human body of smoking they have learned today.

Ask: *Imagine you were asked to try a cigarette. What scientific reason could you give for saying no?* Ask them to practise that scenario with a partner, focusing on the science.

EVIDENCE OF LEARNING:

Observe children's responses and presentations. Can they describe the short- and long-term effects of smoking on human health and lifestyle? Do they use scientific vocabulary when describing the health risks of smoking? Do they use scientific facts to support their arguments against smoking?

CROSS-CURRICULAR OPPORTUNITIES:

There are links with Drama in the role-play element of Challenge 3.

Key information:

http://www.cornwallhealthyschools.org/documents/6398_dat_risk_take_pressure.pdf (QCA Unit B Risk-taking and dealing with pressure) is a useful website.

BODY HEALTH

LESSON 9: CAN YOU SPREAD THE HEALTHY WORD?

LESSON SUMMARY:

In this lesson children reflect on their learning throughout the module. By the end of the lesson they have produced a school booklet about the benefits of a healthy lifestyle. This lesson splits the learning from the rest of the module into three parts: diet, exercise and drugs, recognising how these can all help, or indeed hinder, a healthy lifestyle. Children are required to reflect on their learning and present their findings to an audience.

Key vocabulary:

diet, exercise, drugs, lifestyle

Resources:

Access to work produced in previous lessons. Make available a range of presentation tools for children to select from six sheets of A3 plain white paper to collate research; potential to email the headteacher for the Reflect and review session is preferred

National curriculum links:

Recognise the impact of diet, exercise, drugs and lifestyle on the way bodies function

Learning intention:

To reflect and consolidate learning about a healthy lifestyle

Working scientifically links:

Reporting and presenting findings from enquires, including conclusions, causal relationships and explanations of and degree of trust in results, in oral and written forms such as displays and other presentations

Success criteria:

- I can reflect on my learning about a healthy lifestyle to summarise key information.
- I can explain the impact diet, exercise and drugs might have on my lifestyle.
- I can present my learning for a new audience.
- I can communicate my ideas clearly and precisely using scientific language.

EXPLORE:

Explain to children that they are going to have the opportunity to reflect on their learning from the whole module in this lesson. Remind them that they have been learning about diet, exercise and drugs, and how they contribute to a healthy lifestyle.

Ask: *What other aspects of life contribute to a healthy lifestyle?* Some ideas might include rest, sleep, fresh air and lack of stress or happiness.

Give children independent time to look back over any written materials, investigations and photographs from the module to refresh their thinking about the areas of learning.

In a whole class discussion, collect together some key ideas and facts about each area of learning under the following headings: diet, exercise and drugs. You may wish to record some of their key learning on a whiteboard or on three separate pieces of flip chart paper to focus their attention on the broader view of a healthy lifestyle.

ENQUIRE:

Ask: *Who makes up the wider school community?* Answers might include: pupils, parents, teachers, staff, governors, community groups who use the school, children's centres, parents groups and faith groups.

Explain to them that they are going to work in small groups to present information to the wider school community. In Challenge 1 children write about a healthy diet with support. In Challenge 2 they write about the importance of exercise with prompts and in Challenge 3 they write in groups about drugs and their impact on a healthy lifestyle, and decide how to present their findings.

Challenge 1: Children write about a healthy diet with support

Children write about a healthy diet. They work together to divide up the work between them to avoid repetition (adult support may be needed for this) and present their report as two A3 pieces of paper. They select how to present their work; it could include word processing, scientific diagrams, lists, recommendations or short pieces of writing. They should cover information about the main food groups and the impact a balanced diet has on a healthy life.

Ask: *What are the effects on lifestyle of not eating a healthy balanced diet?*

Encourage them to include anything else they feel is appropriate.

Challenge 2: Children write about exercise using prompts

Children write about exercise. They work together to divide up the work between them to avoid repetition. Their combined work is presented at the end of the lesson as two A3 pieces of paper. They select how they present their work; it could include word processing, scientific diagrams, lists, recommendations or short pieces of writing.

Ask: *Have you included: the benefits of regular exercise on general health? The importance of a rapid recovery rate as an indication of fitness? The effects on health of lack of exercise? Can you think of any other information you think is important about exercise?*

Challenge 3: Children work together to write about drugs and decide how to present their findings

The children work together to write about drugs and their impact on a healthy lifestyle. They work together to divide up the work between them to avoid repetition. Their combined work is presented at the end of the lesson as two A3 pieces of paper. They select how they present their work.

Ask: *Have you included the short- and long-term effects of a variety of drugs? Have you included information about drugs that help us as well as harm us? Have you written about how drugs affect lifestyle?*

REFLECT AND REVIEW:

Spend time as a class sharing and reflecting on the information collated, then ask each child to write an email to the headteacher or principal. It should persuade him or her to publish this information on the school website or as a printed document to share with the wider school community.

EVIDENCE OF LEARNING:

Look at children's completed information sheets. Do they include key facts related to the area of lifestyle presented? Have children made accurate use of appropriate scientific vocabulary? Are the sheets suitable for a wider school community audience? Can they explain the impact of a balanced diet, appropriate exercise and safe use of drugs on a healthy lifestyle?

CROSS-CURRICULAR OPPORTUNITIES:

There are links with English in writing for an audience.

BODY HEALTH

 ENRICHMENT LESSON 1: HOW DO ATHLETES KEEP FIT?

LESSON SUMMARY:

In this lesson children look in detail at the diet and training regimes of a range of athletes. By the end of the lesson they have reviewed what they have learned about diet and exercise, and recognise them as important elements of maintaining a healthy lifestyle. They understand the key variables that help prepare athletes for competition.

Key vocabulary:

diet, exercise, lifestyle, athlete, training, regime, intensity, prepare, competition

National curriculum links:

Recognise the impact of diet, exercise, drugs and lifestyle on the way bodies function

Learning intention:

To research key variables that help prepare athletes for competition

Scientific enquiry type:

Finding things out using a wide range of secondary sources of information

Working scientifically links:

Reporting and presenting findings from enquiries, including conclusions, causal relationships and explanations of and degree of trust in results, in oral and written forms such as displays and other presentations

Success criteria:

• I can research an athlete to identify their diet, training plan and lifestyle.

• I can recognise key patterns in their regime that contribute to their fitness.

• I can relate their training to my learning about diet, exercise and lifestyle.

EXPLORE:

Explain to children that they are going to find out about athletes from a range of sports and how their diet, exercise and lifestyle are crucial in their preparation for competition.

Share the video Corinne Yorston – How I keep fit (Video 1), discussing her diet and training plan in the build-up to a competition.

Ask: *Does Corinne do anything that surprised you to prepare for competition? How does her diet compare with yours? How does varying the type of training aid her in competition build-up? What aspects of her lifestyle contribute to her event preparation?*

Ask: *Some managers and trainers give athletes a strict code of conduct when preparing for competition, such as curfews and bans on certain food and alcohol. Why do you think this might be? Do you agree with this?* Establish that children are clear that competition preparation is wider than diet and exercise, and includes broader lifestyle changes that are crucial to success.

Key information:

You may also wish to refer to Big Cat (pearl band) *Becoming an Olympic athlete* by Beth Tweddle to read more about how her career developed.

ENQUIRE:

Explain to children that they are going to be be researching and presenting information in the form of a report about various athletes. The challenges are differentiated by the level of support provided. Children doing Challenge 1 find out about Ellie Simmonds, with some information provided for them. Children doing Challenge 2 look at Tour de France winners with some key information provided. Children doing Challenge 3 research an Olympian or Paralympian of their choice with minimal support.

Challenge 1: Children produce a report using information provided and present their results

Using Ellie Simmonds (Resource sheet 1) to support them, ask the children to spend time reading about the Paralympic swimmer's diet, training regime and lifestyle in readiness for competition. Ask the children to present their report visually, including an eatwell plate. They may choose their own presentation method to share their findings with others.

Key information:

Children should have access to secondary sources to find out more about Ellie Simmonds if they are unfamiliar with her success.

Challenge 2: Children research an athlete with some support and independently work and present their results

Using Tour de France winners since 2006 (Resource sheet 2) to support them, ask the children to select a winner of the Tour de France from 2006–2013 to find out more about. They should have access to secondary sources and research their diet, training regime and lifestyle in the run-up to competing in the Tour de France. They choose their own style to present their report in.

Challenge 3: Children research an athlete independently and present their results

Ask the children to choose an athlete of their choice to research their training, dietary regime and general lifestyle in readiness for competition. Ask them to avoid athletes being investigated by other Challenge groups but to choose one from a sport that interests them. They should have access to secondary sources for further research and should be encouraged to work independently where possible. They present their report visually and select their own method of presentation.

REFLECT AND REVIEW:

Draw upon the training experiences of any pupils in the class who are training as part of a regional or national squad. Ask them to present to the rest of the class their training pattern and any dietary advice they receive. This may be an opportunity to invite a visitor in from the local professional football or rugby team to talk about their diet and training regime in preparation for match day.

If neither of these is possible, ask children to reflect on their own ambitions, sporting or not, and write on a sticky note the inspiration they have gleaned, by researching an athlete during the lesson, about what it takes to get to the top in their field.

EVIDENCE OF LEARNING:

Observe children as they carry out their research and listen carefully to their presentations. Do they use scientific vocabulary in their presentations? Are their ideas supported by research from secondary sources? Are they able to describe the athlete's training plans, diet and lifestyle? Can they identify the key variables that contribute to an athlete's fitness and competition preparation?

CROSS-CURRICULAR OPPORTUNITIES:

There are links with English with the elements of autobiographies and report writing.

BODY HEALTH

 ## ENRICHMENT LESSON 2: WHAT HAPPENS WHEN ATHLETES CHEAT?

LESSON SUMMARY:

During this lesson children explore why and how athletes use performance-enhancing drugs to enhance their performance. In particular, they look at some high-profile cases of the use of banned substances. By the end of the lesson they have looked at the reasons given by athletes for this drug use and explored the links with what they have learned about keeping their bodies healthy.

Key vocabulary:

doping, cheating, drugs, testing, sport, athlete, performance-enhancing, steroids

Resources:

Access to secondary sources, including the internet, for further research

National curriculum links:

Recognise the impact of diet, exercise, drugs and lifestyle on the way bodies function

Learning intention:

To identify variables that contribute to the way in which athletes cheat

Scientific enquiry type:

Finding things out using a wide range of secondary sources of information

Working scientifically links:

Reporting and presenting findings from enquiries, including conclusions, causal relationships and explanations of and degree of trust in results, in oral and written forms such as displays and presentations

Success criteria:

- I can identify reasons for using performance-enhancing drugs in sport.
- I can present reasons given by athletes when caught cheating.
- I can discuss the role scientific testing plays in identifying cheating athletes.

EXPLORE:

Spend 2 or 3 minutes at the start of the lesson by asking the class to talk to a partner: *What does cheating look like? In what ways might athletes cheat?* Share discussion and feedback as a class and accept all answers at this point.

Share Why do athletes cheat? (Slideshow 1), which features examples of sporting athletes who have tested positive for banned substances and their explanations for why it was in their system. Ask them to pause after each slide and give time to ask questions about anything they do not understand. Establish that children are clear about the potential benefits to the athletes' performance of each drug mentioned in the slideshow:

Testosterone: a type of steroid that gives you additional muscular strength, allowing you to train harder and longer

Nandralone: a type of steroid that increases muscle growth, red blood cell protection and bone density

Cortisone: a steroidal hormone that can give short-term pain relief and reduce swelling, allowing you to train or compete through injury

ENQUIRE:

Explain that each Challenge group is going to investigate more about different ways in which athletes cheat. The challenges are differentiated by the level of support provided. The Challenge 1 group researches Scottish skier Alain Baxter with provided information. The Challenge 2 group researches Lance Armstrong with a provided CV. The Challenge 3 group prepares a judge and jury trial for a case of their choosing.

Challenge 1: Children read provided information about an athlete and write a letter to him

The children work in pairs or as an adult-supported group. Give the children the information about skier Alain Baxter (Resource sheet 1). After they have had the opportunity to read his story, give them time to talk in pairs.

Ask: *Do you think it was fair that he had his medal removed? Did he cheat deliberately? Do you think the athlete liability rules are fair? What were the potential benefits to his performance from taking the drug?*

Ask them to work independently to write a letter to Alain Baxter either in support of him or in support of the appeal hearing findings. Insist that they differentiate between opinion and scientific factual language in their letter.

Challenge 2: Children research an athlete using some provided information and some independent research, and produce a written report

The children research a high-profile case of cheating. Explain to them that they are going to find out more about one of the most well-known cheats in sporting history. Share the Lance Armstrong timeline (Resource sheet 2), which includes personal facts such as his return to cycling after his battle with cancer and his founding of the Livestrong foundation, and, more recently, his admission of systematic doping for at least six years of his career and being stripped of all seven Tour de France titles.

Ask: *What did Armstrong gain physically by cheating? What did he have to lose by cheating? How has public opinion changed over time towards Armstrong? What damage has been done to the reputation of cycling and athletes in general as a result of these revelations?*

Ask them to research further information if they wish to help them to write an opinion piece for the school or local or national newspaper of their views of this particular case. Ask them to include scientific facts where possible and that they need to carefully separate facts, such as dates, times and places, from opinion.

Ask: *Why was he never caught? Who else benefitted from his cheating? What did they have to gain? How did science contribute to the cheating?*

Challenge 3: Children prepare a trial in a drugs test case of their choosing

The children work in pairs or small groups of no more than four. They select an athlete who has failed a drugs test, where they believe that the scientific testing procedure is at fault, to research further. These procedures may include tampering with results or failure to store samples correctly. They may wish to use one of the athletes used in Why do athletes cheat? (Slideshow 1) or others they have heard of or come across in their research. They should put together a case both for and against the athlete in preparation for a judge and jury session in the Reflect and review session. They should take great care to ensure the scientific facts of the case are clear in their presentations, as well as the potential performance-enhancing effects of the drug in question, and be clear about any opinions when they are included.

REFLECT AND REVIEW:

In turns, groups from Challenge 3 present the case for their athlete and the case against them. They should focus on facts they have discovered and include opinions of the athlete and their friends and family, clearly stating the difference between each one. The rest of the class act as the jury, with either an adult or a child presiding over events as the judge. A guilty or not guilty verdict should be decided upon.

EVIDENCE OF LEARNING:

Look carefully at children's presentations. Do they make clear the difference between scientific fact and personal opinion? Can they explain clearly the performance benefits of using a particular drug? Do they include the role scientific testing plays in recognising sporting cheats?

CROSS-CURRICULAR OPPORTUNITIES:

There are links with English in the letter writing, drama and report writing activities.

EVERYTHING CHANGES

INTRODUCTION

This is a challenging module in which children build on their knowledge of living things and how they are adapted to particular environments. They are introduced to the idea that variation in organisms can result in the species becoming better adapted to its environment and that the process of natural selection, over a long period of time, leads to evolution. Although children may have been introduced to the concept of adaptation during their time at school, natural selection and evolution will not have been formally discussed at school prior to this unit. Children learn about how inherited characteristics are passed on from parents to offspring and that environmental variables also affect how organisms look and behave. They explore the process of selective breeding, through which humans can select particular characteristics in different plants and animals to meet specific requirements. They also explore how those individuals in a population that are best adapted to the environment are more likely to live long enough to reproduce and so maintain the population and the survival of the species.

Children learn that it is a combination of inherited characteristics and the effect of environmental variables that ultimately mould the appearance and behaviour of living things through the process of natural selection. Children analyse fossil records, which show that organisms have changed over millions of years and that many have become extinct. Fossils provide evidence for natural selection and evolution. If appropriate, children may complete the module by using their knowledge of natural selection to explain the process of speciation, through which one or more populations of the same species can become separated and change over time to become different species. In this module children carry out investigations to measure the variation between individual organisms of the same species, model the process of dog breeding by selecting parents that have the desired characteristics for producing useful offspring, and design their own animal to suit a specific environment.

When working scientifically, children take measurements to record variation in plants and animals; they use scientific models to describe complex processes such as selective breeding and natural selection, they question themselves and their peers on aspects of adaptation, and they develop their skills for evaluating evidence. Throughout the module children present their work in a variety of ways and have several opportunities for peer assessment and feedback on the work of other children.

Key vocabulary:

population, variation, environment, inheritance, adaptation, selective breeding, generation, survival, natural selection, evolution, fossils, genes, genetics, DNA, extinct, extinction, speciation, question, investigation, fair test, change, measure, predict, prediction, explanation, observations, draw conclusions

FACT FILE:

Through sexual reproduction living things produce offspring that are similar to but not exactly the same as the parents. The offspring are also not identical to each other; even 'identical twins' show slight differences. Each individual has some characteristics of its father, some of its mother and some which appear to be from neither parent. Genes, which are composed of DNA, carry the information that leads to the different characteristics. Each individual gets half of its genes from its male parent and half from its female parent. The variation between individuals occurs because of the different combinations of genes each individual acquires at fertilisation. It is important to note that this variation occurs regardless of the environment in which the organism finds itself.

Humans have been able to use their knowledge of how natural variation occurs to carry out selective breeding in many different types of organism, including food crops such as wheat and apples, animals such as cattle and horses, and pets of different types.

Organisms are also affected by the environments in which they live. The impact of environmental variables can cause variation in the ways in which living things grow, for example, if an animal cannot get much food it is likely to be smaller than if it had access to more food. If food is very scarce then the animal will not survive. Other variables that affect survival include physical and chemical requirements (nutrients, shelter, space, light, etc.) and the likelihood of being eaten (for example, if there are too many predators).

To be able to live, grow and eventually reproduce, individuals in a population are in constant competition with other individuals of the same species as well as with individuals of other species. Organisms that display characteristics that make them better adapted to the environment are more likely to survive and reproduce. These organisms pass their characteristics on to the next generation and so increase the proportion of individuals in the population that have the same beneficial characteristics. Conversely, individual organisms that are not well adapted tend not to survive long enough to reproduce, so the proportion of individuals with these characteristics in a population tends to decrease. Over many generations more and more of the organisms become better adapted to their current environment. This process is called natural selection, and over thousands of years can lead to changes in organisms and ultimately to the evolution of new species.

If the environment changes significantly (either slowly or rapidly) the size of the populations of each species will vary depending on how well the organisms can cope with the new conditions or find another suitable environment. In extreme cases the population decreases in size, as fewer and fewer individuals have the characteristics necessary for survival, and eventually the species becomes extinct. The fossil record provides evidence that over millions of years living things of many types have lived on the Earth and that many have changed. This process of change is called evolution, and natural selection is the scientifically accepted explanation of how evolution comes about. It was the studies of Charles Darwin and Alfred Russell Wallace that first proposed the theory of evolution through the mechanism of natural selection.

When a population of single species is separated into different groups by a geographical barrier, and each group is subject to different changes in their environment, they can evolve differently over many generations to the extent that the individuals from one group cannot breed with individuals from the other group. As a result, two new species are established through a process called speciation. It is because of speciation that the number of species increases, creating the variety of species on the planet. Over millions of years some species survive better than others and many have become extinct. It is appropriate to emphasise that evolution is not a sequential process, but occurs in stages that lead to one species diverging into two different species. So it is incorrect to say chimps evolved into humans. Rather, it is correct to say that a common ancestor evolved into two distinct species (chimps and humans).

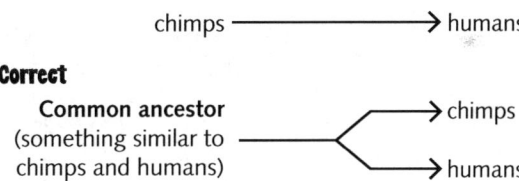

Not correct

chimps ——————————→ humans

Correct

Common ancestor
(something similar to ——————⟨ → chimps
chimps and humans)   → humans

Common misconceptions:

Children may think that only animals adapt to their environment, whereas it is correct to say that all living things adapt.

Children may think that organisms adapt because they need/want to. Plants and animals do not make any decisions it is the individuals that happen to be well adapted that survive and it is these individuals that pass on to their offspring the characteristics that make them more suited to the environment.

Children may say that fossils are the remains of dead plants and animals. It is correct to say that fossils are an imprint of the space left behind when a dead plant or animal decomposes.

Big Cat book links

Evolution Steve Alton 978-0-00-733633-3 Band 14 Ruby	Find out about Charles Darwin's theory of evolution.
Charles Darwin and Alfred Russel Wallace Anna Claybourne 978-0-00-753014-4 Band 18 Pearl	Darwin and Wallace both wanted to find out more about nature. They both travelled around the world, and each discovered new kinds of plants and animals. The two men didn't know each other very well, so how did they come up with the same big idea – evolution?

EVERYTHING CHANGES

 ## LESSON 1: WHY DO LIVING THINGS VARY?

LESSON SUMMARY

In this lesson children investigate and discuss how characteristics of living things, for example, height, size or colour, vary from individual to individual. By the end of this lesson children are able to recognise that characteristics in living things can be inherited, affected by the environment, or result from a combination of the two.

Key vocabulary:

variation, characteristic, environment, inherited, measurement, data, compare and contrast

Resources:

Rulers, metre sticks or tape measures, sticky notes, large sheets of paper (A3), access to the internet or books for further research

Health and safety:

Discussions about inherited characteristics in humans should be handled sensitively as some children may not be aware of their biological parents. Children should ask permission before measuring each other.

Key information:

Most characteristics of living things are inherited, such as eye colour, natural hair colour, or the shape of the face, but some characteristics are influenced by the environment, for example, for humans the language we speak or whether we have scars. Inherited characteristics are passed on from parents to offspring by genes, while those caused by the environment during their life cannot be passed on. The relative impact of inheritance and the environment is still an area of scientific debate for many characteristics, as some, for example, behaviour, intelligence, body mass and height, are caused by a combination of inherited and environmental variables.

National curriculum links:

Recognise that living things produce offspring of the same kind, but that offspring normally vary and are not identical to their parents

Working scientifically links:

Recording data and results of increasing complexity using scientific diagrams and labels, classification keys, tables, scatter graphs, and/or bar and line graphs

Learning intention:

To identify ways in which living things of the same kind vary and to begin to think about why these variations exist

Scientific enquiry type:

Grouping and classifying things

Success criteria:

- I can identify key features of living things.
- I can describe ways in which characteristics of living things may vary.
- I can suggest reasons for the variation between different living things.

EXPLORE:

Organise children into pairs for a structured discussion. Display slide 1 of the Dogs slideshow (Slideshow 1), which shows a picture of a golden retriever and a dalmatian. Ask pairs of children to identify the similarities and differences between the two dogs by completing the Dogs compare and contrast grid (Resource sheet 1). Children may find it hard to comment on some of the variables for comparison (for example, diet, running speed) and therefore access to research materials such as the internet and textbooks should be provided to support this.

Ask: *Why do you think these animals look like they do? How are they similar and how are they different?*

Discuss reasons for the similarities and differences. Collect children's ideas on sticky notes, encouraging suggestions such as 'because of their parents', 'because of how much they eat', 'because of the way the owner looks after them' (for example, brushing the coat), 'because of how much exercise they get'.

Label two large sheets of paper with the headings 'inheritance' and 'environment' and explain these big ideas using humans as the example.

Ask: *Which features do you think are inherited? Which are caused by the environment? Do any of your features fit under both headings?*

To complete the Explore activity, ask children to add a key to their compare and contrast grid and to use different colours to show which characteristics of the two dogs are inherited, environmental or both. Children should be reminded that although there are many differences between the dalmation and the golden retriever, they are both still dogs.

ENQUIRE:

Start by linking this Enquire section of the lesson to what children have already learned by reminding them of the many variations between different types of dogs. Explain that variations in characteristics occur in different individuals of the same kinds of plants and animals as well, and that they are going to investigate variation in living things further by looking more closely at humans and comparing their findings. Children should work in groups of three or four.

The challenges are presented on the Challenge slides to be displayed on the board, or printed out and placed in the centre of the table.

Challenge 1: Children observe variations between two humans

Give each group of children the Humans compare and contrast grid (Resource sheet 2) to support them as they observe differences between two humans and record their observations. Then, either give the children two humans to compare from the range of images provided in the Humans photo slideshow (Slideshow 2), or ask them to compare two of their peers. Start the children off by filling in the first three comparisons for them.

Encourage them to discuss each feature as they make their comparisons.

Ask: *Is the feature you are discussing the same for every human or does it vary sometimes? Where do you think this feature came from? What makes a feature inherited? What makes a feature environmental? Are any features inherited and environmental?*

When the children have completed the compare and contrast grid ask them to highlight which variables are inherited, environmental or involve a combination of both, as they did in the Explore activity.

Challenge 2: Children observe variations in groups of humans

Ask the children to select groups of humans from the Groups of humans photo slideshow (Slideshow 3) and produce their own compare and contrast grid, following the structure from the Explore activity. Encourage the children to compare features that they think are important when observing human characteristics. The children should complete their grid and then add highlighting to show which features are inherited, which are environmental and which are a combination of both.

Ask: *Is it difficult to label any of the features? If so, why? Which features do you think are inherited? Why? Which features are affected by the environment? Is being tall inherited or environmental? Is being good at sport inherited or environmental? Is being well behaved or naughty inherited or environmental?*

Challenge 3: Children measure variation among humans

Ask the children to select a number of characteristics that they think vary in humans, for example, height, weight, arm span, shoe size, eye colour and hair colour. Ask them to conduct a survey of their class, taking measurements where appropriate, in order to show the variation in these characteristics that occurs among their classmates. The children should record their results in a table and present them as histograms or graphs. Explain to them that how they present their findings graphically depends on which characteristics they chose. Some, for example eye colour or shoe size, are discrete variations with a limited number of possibilities and the children record how many individuals have a particular eye colour or a particular shoe size, for example. Other variations, for example height or arm length, are continuous and could be any value for an individual.

Remind the children to consider which characteristics are inherited, which are environmental and which are a combination of both.

Ask: *Which variations were easy to distinguish? Which were more difficult? Do you think a particular characteristic, for example eye colour, is inherited? What makes you say that? Which characteristics do you think are caused by the environment? Which characteristics do you think are inherited but can be influenced by the environment?*

Key information:

The terms 'inherited' and 'environmental' are important for the rest of the topic so children should be encouraged to use these terms correctly during discussions.

REFLECT AND REVIEW:

Give children the Plants and animals photo slideshow (Slideshow 4), or project the photographs onto the whiteboard. Ask them to work in small groups to list the key features of each plant and animal, and decide which of the features are inherited (passed on by genes from parents to offspring) and which might have been caused by environmental variables or conditions.

EVIDENCE OF LEARNING:

Observe and listen to children's conversations during the compare and contrast tasks. Are children able to identify features of living things that show variation from individual to individual? Can children suggest what might have caused the variation? Do they use the terms 'inherited' and 'environmental' correctly? Can children correctly name some features that are inherited, some that are the result of the environment and some that result from both inherited and environment influences, such as weight (as it is affected by inheritance and diet) or height (as it is affected by inheritance and diet)?

EVERYTHING CHANGES

 ## LESSON 2: CAN YOU BREED A DOG FOR A SPECIFIC PURPOSE?

Key vocabulary:

variation, breeding, inheritance, offspring, characteristics, crossbreed, generation

Resources:

Secondary sources of information for further research

LESSON SUMMARY

In this lesson children develop their understanding of inheritance and explore how characteristics are passed on from parents to offspring. They sort dogs into breeding pairs in order to produce offspring with particular characteristics. Some children extend these ideas to plants. By the end of this lesson children are able to explain that many characteristics are inherited and that inheritance plays a major role in variation.

National curriculum links:

Recognise that living things produce offspring of the same kind, but that offspring normally vary and are not identical to their parents

Working scientifically links:

Identifying scientific evidence that has been used to support or refute ideas or arguments

Learning intention:

To recognise how organisms can be bred to select particular characteristics in their offspring

Scientific enquiry type:

Finding things out using a wide range of secondary sources of information

Success criteria:

- I can give an example to describe how organisms might be bred together to produce offspring with particular characteristics.
- I can use the terms 'inheritance' and 'characteristics' correctly when talking about breeding and offspring.
- I can understand that the characteristics of the offspring come from the parents.

Key information:

Children may not have heard the term 'breeding' before and this may need to be explained. Breeding is the process of sexual reproduction by which living things produce offspring.

Key information:

At this stage children are not expected to understand how genes and chromosomes work, but they should be aware that genes contain the information that determines the characteristics of living things. It is acceptable to simply refer to the process of characteristics being passed from parents to offspring as 'inheritance'.

EXPLORE:

Display slide 1 of the Dogs and their offspring slideshow (Slideshow 1), which shows photographs of a labrador and a poodle side by side. Ask children what would happen if the two dogs were bred together.

Discuss with children what they think the offspring of a labrador and a poodle would look like and then display slide 2 from Slideshow 1, which shows a labradoodle. Explain that the labradoodle is the offspring of a labrador and a poodle, and that this crossbreed has inherited characteristics from both its parents. Explain that one of the reasons that the labradoodle was bred was to combine certain characteristics of each parent. The poodle does not shed much fur and labradors are easy to train, so the idea was that the crossbreed could be used as a guide dog for people who were allergic to dog fur.

Next, give children the Dog breeding information cards (Resource sheet 1) of the Jack Russell terrier, greyhound, Siberian husky and samoyed.

Explain to them that a family you know wants a small friendly dog (under 40 cm tall) that they can have as a pet for their children. They would like a white dog with white fur that will be welcoming to strangers and can live outside. Ask children to look at their cards and discuss, in small groups of two to four, which two dogs would be best to breed together to make the right dog for the family.

ENQUIRE:

Explain to children that humans have been selectively breeding dogs for hundreds of years and so dogs are a useful example for investigating inheritance. Explain to children that they are going to look at the ways in which new breeds of dogs could be developed to match particular needs.

The challenges build on the Explore activity, and are differentiated by the level of support and the extent to which children can apply their understanding in another context. In Challenge 1 children are supported in making their decisions by limiting the number of 'dog requirements' that they are given or the number of dogs that they can choose from. In Challenge 2 children are given limited help as they match dogs to create a set of desired features, while Challenge 3 asks children to extend their understanding of inheritance and breeding to plants. Children should work in groups of three or four.

Key information:

Check that children understand that one parent is female and one is male.

Key information:

Some children may use the word 'generation' when describing selective breeding, for example, we are breeding the second generation to produce the third generation. This vocabulary should be encouraged if used correctly.

Key information:

Although the topic of inbreeding is not included in the national curriculum, there is an opportunity here to discuss the disadvantages of inbreeding in dogs and the physical defects or illnesses that can result in dogs that are continually inbred.

Key information:

Challenge 3 introduces ideas that will be revisited and built upon in the next lesson, Lesson 3 (How can we make our food better?) and in Lesson 10 (How does natural selection work?), although children not completing Challenge 3 in this lesson will not be at a disadvantage.

Challenge 1: Children decide on which two breeds of dog to cross to match one set of requirements

Organise the children into small groups and give each group the full set of Dog breeding information cards (Resource sheet 1) together with a list of Requirements for dogs (Resource sheet 2), which describes the features that different people are looking for in a dog.

Ask: *Why have you chosen those dogs to breed?*

Challenge 2: Children choose pairs of dogs to breed to meet the requirements of different people or families

Give each group of children the full set of Dog breeding information cards (Resource sheet 1) together with a list of Requirements for dogs (Resource sheet 2), which describes the features that different people are looking for in a dog.

Ask the children to explain their decisions and to consider any problems that may arise when breeding two dogs with very different characteristics.

As the children become more confident, ask one child to describe a dog with characteristics taken from the requirements cards and then ask the other children in the group to suggest what the characteristics of the parents would need to be if they wanted to breed a dog with those requirements.

Ask: *What characteristics do you want the offspring to have? What characteristics do you think you will get from each parent dog? Do you think you will always get the characteristics that you want in the offspring? Could there be any disadvantages to breeding the offspring again and again?*

Challenge 3: Children apply the idea of breeding to plants

Give each group of children the Apple breeding chart (Resource sheet 3), which includes a table of apple characteristics. Ask the children to answer the questions on the chart about which apples they would cross together to produce apples with particular characteristics.

Ask: *What characteristics did you want from each parent? Do you think you will always get the characteristics in the offspring? What might be the advantages and disadvantages of breeding plants?*

Provide the opportunity for the children to use secondary sources to find examples of other food plants that have been bred to either produce more or taste better.

REFLECT AND REVIEW:

Display the Wolf and dogs chart (Slideshow 2). Explain to children that all the different types of dogs we see today have been bred from the wolf.

Ask: *How do you think it has been possible to get all these different types of dogs? How long do you think it has taken?*

Next, display the Wild mustard chart (Slideshow 3) and explain to the children that many different vegetables, including cabbage, broccoli, cauliflower, kale and kohirabi, have all been bred from the wild mustard plant.

Ask: *How do you think it has been possible to get all these different types of vegetables? How long do you think it has taken?*

Explain to children that you would like them to think about some questions in preparation for the next lesson (Lesson 3).

Ask: *Why might humans want to control breeding in animals? Why might humans want to control breeding in plants? What problems might this cause?*

EVIDENCE OF LEARNING:

Observe and listen to children's conversations during the dog breeding task. Do children understand that nearly all of an offspring's characteristics come from the parents? Can they use the term 'offspring' correctly? Can children select appropriate parents to produce offspring that suit each requirement? Can children select more complicated breeding choices to try to produce even more suitable offspring? Can they explain the benefits of cross breeding? Are children completing Challenge 3 able to recognise the ways in which they could create new types of plants with particular inherited characteristics? Do children use the terms 'inherited', 'characteristics', 'offspring' and 'breeding' correctly?

EVERYTHING CHANGES

 ## LESSON 3: HOW CAN WE MAKE OUR FOOD BETTER?

LESSON SUMMARY

In this lesson children build on the selective breeding activities in Lesson 2 and extend their learning to the subject of selective breeding for food, and its advantages and disadvantages. By the end of this lesson children understand how selective breeding works and are able to describe some of the ways in which selective breeding can benefit humans and why some people disagree with selectively breeding animals.

Key vocabulary:

variation, selective breeding, inheritance, offspring, characteristics, generation

Resources:

Secondary sources of information for further research

National curriculum links:

Recognise that living things produce offspring of the same kind, but that normally offspring vary and are not identical to their parents

Working scientifically links:

Identifying scientific evidence that has been used to support or refute ideas or arguments

Learning intentions:

To describe selective breeding and evaluate different people's opinions

Scientific enquiry type:

Finding things out using a wide range of secondary sources of information

Success criteria:

- I can describe selective breeding.
- I can describe some of the ways in which selective breeding can benefit people.
- I can explain why some people do not agree with selective breeding.
- I can prepare an argument for or against a particular viewpoint.

EXPLORE:

Remind children of the wide variety of foods that we eat.

Ask: *Where does food come from? Does all the food you have eaten exist in the wild or have we made it ourselves? How have we made it?*

Display images from the Wild to bred slideshow (Slideshow 1), which shows photographs of plants and animals alongside farmed versions, for example, a wild boar alongside a farmed pig. Ask children to suggest ways in which humans have changed how these living things look. Remind children of the dog breeding and apple activity from Lesson 2.

Next, show children the Selective breeding video (Video 1), which shows how selective breeding of cows is carried out on a farm to produce offspring with more muscle for meat.

Ask: *If two muscular cows breed will all their offspring be as muscular as each other? Why does the farmer have to breed the offspring many times to get the best cows?*

It is important when talking to children about their answers to note that not all offspring inherit the same characteristics and that some offspring are more muscular than others. Selective breeding is not a quick process, as it can take many generations of breeding to make significantly improved offspring. Emphasise the term 'selective breeding' so children are clear about its meaning.

Key information:

This lesson focuses on traditional methods of selective breeding. It does not cover the more recent and controversial development of technologies for producing genetically modified (GM) foods.

ENQUIRE:

Inform children that they are going to take part in a debate about the advantages and disadvantages of selective breeding. They are each going to be given a role. (You may wish to allocate roles on the basis of the comments or observations that children have made during the first part of the lesson.) Differentiation will be by the role children play within a group, as explained in the challenge summaries.

Organise children into six groups (of three to four) and give each group one of the six Debate character cards (Resource sheet 1). Ask them to look at their card and prepare, using secondary sources, what they are going to talk about to the class.

Explain that someone from each group is chosen to say who they are and speak for about two minutes on what they think about selective breeding. Each of the other groups has a minute to come up with a question to ask that person/group. They then ask that question and member(s) of the speaking group can respond.

After the questions, every group should write down the main argument put forward by the speaking group on the Debate framework sheet (Resource sheet 2). They should then give that group a mark in the last column.

After all the groups have spoken, each group needs to fill in the last row of the debate sheet, deciding who had the best argument and what the groups think about selective breeding.

The idea is that within each group children take on roles according to the level of challenge that suits them, that is, each group is made up of children working at different challenge levels. In each group there are children who join in the group discussion to develop their case or to decide on questions for other groups; children who can present the case for the group; and children who can think quickly to answer the questions posed, as well as put forward counter arguments and responses showing that they understand the ideas.

Challenge 1: Children take part in discussions within their group

Encourage the children with a basic understanding of the arguments to contribute to the activity by asking questions in their group or during the debate if prompted.

Challenge 2: Children contribute to the presentation

Encourage the children who have a good grasp of the arguments to contribute to presenting the case for their group and to ask relevant questions in the group or debate parts of the lesson.

Challenge 3: Children contribute to a for-and-against debate

Encourage the children who have a good understanding of both sides of the debate to answer questions and present relevant counter arguments. They should appreciate how some arguments are stronger than others and that some arguments are driven by different motivations (for example, characters 5 and 6 of the Debate characters cards).

REFLECT AND REVIEW:

After children have had time to complete the Debate framework sheet (Resource sheet 2), ask them to discuss the arguments.

Ask: *Which point in your argument is the strongest? What made it a strong argument point? Were there any characters that didn't have a very strong argument? Do you think the characters you have been told about could ever agree with the other side of the debate? Have you heard any arguments today that changed your mind? Do you think that you might be made to change your mind after today?*

When they have finished their discussion, ask children to take a class vote on which was the best argument.

Explain to children that through selective breeding, humans choose only the 'best' plants and animals to survive. Ask them to think of situations where it is nature that chooses the 'best' plants and animals to survive. Lead the discussion by prompting children with some examples, for example, lack of food or threat of predators.

This final discussion links to the next lesson, Lesson 4, during which children are introduced to the idea that changes in the environment can impact upon animals and plants. In the remaining lessons they explore how this affects their chances of survival.

EVIDENCE OF LEARNING:

Listen to the children's arguments during the debate and their discussions during the debate preparation. Do they understand how selective breeding works? Do they understand that selective breeding happens over many generations and takes a long time? Do they appreciate that selective breeding can happen in animals and plants? Are children able to appreciate and record arguments both supporting and rejecting selective breeding? Are they able to identify and either write or articulate evidence that supports or refutes the arguments?

EVERYTHING CHANGES

 ## LESSON 4: HOW DOES THE ENVIRONMENT AFFECT PLANTS?

LESSON SUMMARY

This is the first part of a two-part lesson. In this lesson children begin to investigate ways in which the environment can affect how plants grow. They make observations, and plan and set up a fair test to investigate a demonstrable effect that the environment has on plants. By the end of the lesson children are able to predict how different environmental conditions might affect the growth of plants. In Lesson 5 children set up their tests, and analyse and explain their results.

Key vocabulary:

population, variation, environment, observation, variable

Resources:

Access to wild plants in different habitats or photographs of wild plants in different habitats

Health and safety:

If possible, children should observe wild plants in different habitats. When outside, ensure children's safety and remind children that they should observe the plants, not touch them. Some plants may sting children or cause a rash.

National curriculum links:

Identify how animals and plants are adapted to suit their environment in different ways and that adaptation may lead to evolution

Working scientifically links:

Planning different types of scientific enquiries to answer questions, including recognising and controlling variables

Learning intention:

To observe the effects of the environment on plants and design an experiment to investigate some of these effects

Scientific enquiry type:

Carrying out comparative and fair tests

Success criteria:

- I can observe how the environment affects plants in nature.
- I can plan a fair experiment to investigate ways in which the environment might affect plants.
- I can give examples of how the environment affects plants in nature.

EXPLORE:

Organise the class into small groups of up to four children. Remind children of the requirements of a plant. This is a topic that they should have covered in Year 3, Module 1. Encourage children to think of other things that might affect how well plants grow in the wild.

Give children a few minutes to complete the card match, then ask them to share their matches and record their ideas on the board. Alternatively, use the Plant match interactive (Interactive 1). Ask children to explain their choices, for example, why the roots should be matched with rain.

To link this activity to the Enquire activity, display slide 1 of the Comparing leaves slideshow (Slideshow 1), which shows images of leaves on a plant that have been in the sunlight and leaves from the same type of plant that have been in the shade.

Ask: *Why are there differences between the leaves? What are the variables that might affect them?*

Key information:

If children struggle to explain the differences between the leaves in the sunlight and those in the shade, prompt them to first think back to why plants have leaves and then lead them to the idea that in the shade the plant has less light and so needs bigger leaves to collect enough light to survive.

ENQUIRE:

Explain to children that their challenge is to explore the local area for evidence of environmental impact on plants.

Ask: *What do you expect to notice? What types of differences might we see between individual plants of the same kind? What do you think might cause these differences? How many different kinds of features can we see on outdoor plants that are linked to the environment?*

Following the outdoor exploration, children discuss their observations and plan a fair test to investigate the impact that the environment can have on plants. At this stage you might wish to let children know what plants and materials are available to them in the next lesson. The challenges are differentiated by the amount of support provided in the planning and execution of the fair test and by the data presentation requirements. Children should work in groups of three or four.

Challenge 1: Children explore the local area and observe the growth of wild plants

Explain to the children that their task is to look for examples of variations in plant populations which could be caused by the environment. If an outdoor activity is inappropriate or not possible, show children the Plant environment clip (Video 1) and ask them to comment on their observations. Children should record (by drawings, notes and/or photographs if outside) some of the effects that they observe, to share during the Reflect and review part of the lesson.

Ask: *How are the plants in the population different? What kinds of conditions did you find (does the video show) the plant in, for example, light or dark? Can you see a link between how the plant looks and the conditions it's growing in?*

Explain to the children that their task is to identify one environmental variable and to design an investigation to see what effect this variable has on plants. They can use the planning template, My observations (Resource sheet 1), to help them devise their investigation.

Challenge 2: Children explore the local area, and observe and record the growth of wild plants, noting variations in populations

Ask the children to make and record their observations about the effects of the environment on plant growth, using appropriate methods, such as taking notes, drawing diagrams and/or taking photographs. They share these observations in the Reflect and review activity.

Ask: *What effects did you notice that the environment has on the plants outside/on the video? What did you observe about the plants?*

Encourage the children to discuss their observations within their group and then to design an investigation to test the effect of an environmental variable on plants.

Ask: *What environmental variable do you want to investigate? Which dependent variable will you measure in the plants? How will you do this? What should you keep the same in your experiment to make sure the test is fair? What do you think the result of your investigation may be?*

Challenge 3: Children explore the local area and observe the growth of wild plants, noting any examples of variation in plant populations

Ask the children to record (by making drawings or taking notes or photographs) some of the effects of the environment on plants that they observe during their outside activity or on the video. They share their findings during the Reflect and review part of the lesson.

Ask: *How many different ways did you see that the environment has affected the plants? What differences did you see between the individual plants? Do you think that all the differences between the plants are caused by environmental variables? What else might be involved?*

Allow time for a discussion with the children on their observations, then ask them to design an investigation to test the effect of an environmental variable on plants.

Ask: *What independent variable and dependent variable do you wish to investigate? Why did you choose these variables? Did you see something outside/on the video that gave you the idea? How will you make sure that your test is fair? How will you record and present your results? Do you need to use a type of graph?*

REFLECT AND REVIEW:

Ask children to share the photographs and other records that they collected during their exploration of the local area (or watching the video) with the rest of the class. Ask them to describe the features of the plant that interests them (and to show in a photograph, if possible) and to describe the environmental variable that has caused this effect, for example "This photograph shows some plants getting taller the closer they are to a tree. The variable that causes this is shade and the plant's need for sunlight."

EVIDENCE OF LEARNING:

Observe children as they make their observations and record what they see. Are they able to successfully identify differences within plant populations? Can they suggest what environmental variables caused those differences? Can children design a suitable investigation?

Key information:

Some children may begin to explain the reasons behind the differences seen in the plants, for example, the reason the plants grow taller is because they need to reach sunlight. These comments should be noted and revisited during discussions centred on the explanations for plant features, which takes place in the next lesson.

EVERYTHING CHANGES

 ## LESSON 5: HOW DO ENVIRONMENTAL VARIABLES AFFECT PLANTS?

LESSON SUMMARY

This is the second part of a two-part lesson. In this lesson children carry out and analyse the results of the investigations they planned in Lesson 4. By the end of the lesson children are able to describe the relationship between different environmental conditions and variations in plants. It is advisable to divide this lesson into two shorter sessions; the first to set up the fair test and the second to interpret the results.

Preparation required:

Prior to this lesson children's plans for their investigation should be reviewed and the plants, seeds and equipment collected. The fair tests may be kept very open so that children are able to investigate a wide range of variables or, if appropriate, they can be restricted by using the same type of plants/ seeds and/or the number of variables. The challenges in this lesson exemplify one of many possible fair test investigations.

Key vocabulary:

population, variation, environment, observation, variable, respond

Resources:

Petri dishes, cotton wool, cress and mustard seeds, dark paper

Health and safety:

Remind children not to consume the cress or cress seeds. Make reference to SAFETY CODE using plants (Be Safe! section 9).

National curriculum links:

Identify how animals and plants are adapted to suit their environment in different ways and that adaptation may lead to evolution

Working scientifically links:

Planning different types of scientific enquiries to answer questions, including recognising and controlling variables

Learning intention:

To investigate the effect of environmental variables on plants and interpret the results

Scientific enquiry type:

Carrying out comparative and fair tests

Success criteria:

- I can conduct a fair test to investigate the effect of an environmental variable on the growth of plants.
- I can record and measure ways in which some plants vary in their response to an environmental variable.
- I can suggest reasons why some plants in nature respond in different ways to environmental variables.

EXPLORE:

Ask children to list all the environmental variables they can think of that affect plants. Remind them of the work they did in Lesson 4 and ask them to look over the predictions that they made when planning their investigations.

Ask children to individually read aloud their variables and then ask the rest of the class to raise their hands to show if they agree or disagree that a variable affects plant growth.

Ask: *Do you think that all the plants will respond in the same way to this variable? Will they all grow to the same size? Will they all germinate?*

ENQUIRE:

Explain to children that they are going to carry out their investigation, record their results and propose a conclusion that is supported by their findings. Provide children with cress seeds, petri dishes and cotton wool to grow the seeds in. They should set up two batches of cress and change one environmental variable, for example amount of light or water, for one of the batches. Everything else should be kept the same to make it a fair test. It is advisable to use a set number of seeds (20–50) in each batch and to space them out over the cotton wool. They should also record when a plant dies. The investigation could be run until all the plants have died or terminated after 7–10 days.

The challenges are differentiated according to the level of detail that is expected in children's results and interpretation.

Challenge 1: Children set up a fair test investigation

Ask the children to check that they still think their planned investigation is fair and then to set up the materials, as described in the Enquire part of the lesson. Give the children the extended writing frame, My observations (Resource sheet 1), and ask them to record why they made the prediction they did.

Ask: *What effect do you think the environmental variable you have chosen will have on the plants? Why? What other variables do you think will affect the growth of cress? Are all the plants affected in the same way?*

Explain to the children that they should record what they observe as the cress seeds germinate and grow, in a table. A template is provided in the Extended writing frame, My observations (Resource sheet 1).

Challenge 2: Children set up a fair test investigation to justify their predictions and explain their results

Ask the children to review the plan of their investigation from Lesson 4. Ask them to make sure that their plan includes an independent variable to change, a dependent variable to measure, a list of control variables to keep the same, a prediction and the changes that they have chosen to monitor. Ask the children to justify their prediction, show how they will record their results and explain why their results will help them to decide whether their prediction is correct or not.

Explain to the children that they need to monitor the growth of the cress over time and record the changes that they have decided to monitor. These might include the number of cress seeds that germinate each day, the size of individual plants as they grow or the length of time each plant survives.

Ask: *What effect do you think the environmental variable you have chosen will have on the plants? Why? What other variables do you think will affect the growth of cress? Are all the plants affected in the same way? What do we need to remember when planning a fair test? What do you think would happen if you used a different kind of plant? Would you get the same results?*

Challenge 3: Children set up fair tests to compare the effects of an environmental variable on two different plants

Explain to the children that they will be supplied with two different types of seeds, so that they can compare the effect of their environmental variable on two different plants. Ask them to check that their planned investigation is fair and that their plan includes a method, a choice of independent variable to the change, a choice of dependent variable to measure, a list of control variables to keep the same, a prediction, a scientific explanation of the prediction and a results table. The children should also be asked to include an evaluation of their own plan, including why they have chosen to test that environmental variable, what could go wrong with the test and how they could reduce the risk of things going wrong.

Supply the children with cress and mustard seeds. Explain to them that they can plant two batches of cress and two batches of mustard, and change the same environmental variable for each type of seed, while keeping all other variables the same as far as possible.

Ask: *Why have you chosen to test that environmental variable? What effect do you think this variable will have on the plants? Why? What other variables do you think will affect the growth of the seeds? Are all the plants affected in the same way? What do we need to remember when planning a fair test? Is there anything you could do to improve your fair test?*

When the investigation has been completed ask the children to make sure that they have recorded all the information in an appropriate form.

Ask: *Did the two types of plants respond to the environmental variable in the same way? What do you think would happen if you mixed the plants together? What do you think would happen if you used more seeds in the same sized tray so the plants are much closer together? Or in a larger tray so that the plants are more spread out?*

Key information:

Although individual plants of the same type respond in slightly different ways to the environmental variables, different types of plants can respond very differently to the same conditions. This is because the inherited characteristics of different plants (and all living things) are the major variables that determine how well they are suited to the environment.

REFLECT AND REVIEW:

Show children the Plant growth animation (Animation 1), which shows two plants growing in different ways. Ask them to predict how the environment for each plant might be different and to suggest why that could cause the type of growth they see in the animation.

Encourage children to suggest some other environmental variables that might affect how well plants of the same kind within populations grow. Such variables are the overall amount of light, water and nutrients in the soil; or competition between the plants for space, light and nutrients; or how many animals in that environment eat that particular type of plant. Some children may suggest (correctly) that inherited characteristics play an important part in how living things respond to environmental variables. This idea is developed in the remaining lessons of the module.

EVIDENCE OF LEARNING:

Are children able to successfully plan a fair test to show the effects of changing an environmental variable? Can children predict how environmental variables might impact plants? Can children make and explain their prediction? To what extent the children plan their own fair tests. Are they able to evaluate their own plan? Can they apply their knowledge of environmental variables and suggest reasons for the ways in which different plants respond to changes in environmental conditions?

EVERYTHING CHANGES

 ## LESSON 6: HOW DO LIVING THINGS SURVIVE?

LESSON SUMMARY

In this lesson children continue to develop their understanding of the idea that changes in the environment can impact on living things. They examine ways in which the physical features and behaviour of living things make them more suited to the particular habit in which they live, and how adaptations of living things help them to survive in their environment. By the end of the lesson children are able to give examples of how living things can be adapted to their environment and how these adaptations help them to survive.

Key vocabulary:

environment, survival, habitat, temperature, predator, prey, adaptation

Resources:

Large pieces of paper, access to secondary sources of information, including the internet or books, for further research

Information:

The following website contains useful information that can be used to support teaching throughout the lesson: http://www. bbc.co.uk/nature/ adaptations.

National curriculum links:

Identify how animals and plants are adapted to suit their environment in different ways and that adaptation may lead to evolution

Working scientifically links:

Reporting and presenting findings from enquiries, including conclusions, causal relationships and explanations of results, in oral and written forms such as displays and other presentations

Learning intention:

To explore ways in which living things are adapted to suit the environments in which they live and to help them survive

Scientific enquiry type:

Finding things out using a wide range of secondary sources of information

Success criteria:

- I can give examples of how living things are adapted to suit their environment.
- I can suggest how an animal's features and behaviour help it to survive.
- I can give reasons why changes to an environment could affect the survival chances of an animal.

EXPLORE:

Write the following question on the board: "Who lives where and why"?

Organise children into small groups of three or four and hand out the Habitat and environment cards (Resource sheet 1) to each group. Call out the names of a plant or animal and ask children in which one of the habitats on the cards that plant or animal might live. Suggested plants and animals include frog, fish, rabbit, goldfish, hamster, rat, fox, sunflower, cactus and cress, but others can be added as long as both plants and animals are included, and at least one of the habitats featured is suitable for each example.

Ask: How did you decide which plant/animal lived where? What makes this plant/animal suited to living in a particular place?

Encourage children to share their ideas in their groups and to record them on a big piece of paper to share with the class. Ask children to think about the things that organisms need to survive, for example, food, water, suitable habitat and correct temperature. Help them to articulate that although some things may be able to live in more than one type of environment, there is often a particular type of environment that they are particularly suited to.

Ask: Is there anything else you can think of that might make it difficult for living things to survive in a particular place?

Prompt children to think about the risks of being eaten by another organism, for example, or to consider that when too many of the same type of organism live in one place they compete for space, food and/ or water: when things compete, some win and some lose.

Ask: How do plants and animals survive in places where there isn't much food or water? How do living things survive when they are not in a suitable habitat or the temperature is not right for them?

Discuss children's answers and use them to introduce the the idea that pressures and changes in the environment can threaten an animal's survival, for example, extreme cold, lack of food and high levels of competition.

ENQUIRE:

Explain to children that their challenge is to look at some pressures and changes in the environment, and the effects that these changes can have on animals. Children need access to the internet or relevant books so that they can research an animal and its habitat, and present information about what they discover to the class.

The task is differentiated by the amount of detail and explanation required in children's presentations, and the amount of support provided, in particular the support needed to connect the characteristics of the animal to an environment. Make sure that children understand the differences between 'describe' (Challenge 1) and 'explain' (Challenge 2). In order to explain something children must be able to make it clear how an animal's features and behaviour help it to survive. In Challenge 3 children need to to show an understanding of the consequences of environmental change on the animal's ability to live in that environment.

Ask children to choose one animal from a list (for example, polar bear, camel, whale, shark, rabbit or chameleon) and to research how these animals are adapted to survive in their environments.

Challenge 1: Children choose an animal to research and describe how at least one feature of the animal relates to the environment in which it lives

Tell the children to record details about the animal, including as many physical features as possible. Ask them to pick at least one of these features and describe how this feature relates to the environment in which the animal lives.

Ask: *What type of environmental conditions does this animal have to deal with? Does it have any physical features that help it to cope with these conditions and so help it to survive? Does it behave in a certain way that helps it survive?*

Challenge 2: Children choose an animal to research and explain how the animal's physical and behavioural characteristics help it to survive in its environment

Tell them to record information about their chosen animal, including information about the environment in which it lives and the conditions that may occur in the environment. Ask the children to describe the physical and behaviour characteristics of the animal, and explain how these characteristics help the animal to survive.

Ask: *Which features/behaviour help this animal to survive? How do these features/behaviour help this animal to survive, for example, to catch its food or avoid being eaten? How might the characteristics/behaviour of the animal help it to compete more successfully against other animals?*

Challenge 3: Children choose an animal to research, suggest how a change in the environment could cause problems for the animal and identify physical features or behaviour that might need to be different if the animal is to survive in the changed environment

Tell them to record details about the environment in which the animal lives, and the physical features and behaviour of the animal that make it suitable for that habitat. Ask the children to suggest how changes in the environment may cause problems for the animal and what physical or behavioural changes the animal may need to survive in the altered environment.

Ask: *Which features/behaviour help this animal to survive in its environment? How do these features/ behaviour help this animal to survive? What would happen if the environment suddenly changed, for example, the North Pole became hot? Can you find an example of an animal that has changed its behaviour in order to survive?*

REFLECT AND REVIEW:

Give children the opportunity to present the information they have gathered. Provide them with the Presentation feedback table (Resource sheet 2) so that they can provide feedback on other presentations to their classmates.

Ask: *If there were a change in the environment, what might happen to a particular type of animal over a long period of time?*

Explain to children that they are going to consider this question in the next lesson.

EVIDENCE OF LEARNING:

Listen to children's presentations and observe what they write in response to the other presentations. Do children describe the environmental conditions in a certain habitat, for example, the desert is very hot and dry? Can they describe an animal's adaptations to these conditions, for example, some elephant have large ears? Are children able to explain animals' responses to changes in the environment, for example, some elephants' ears have a large surface area which allows heat to be lost quickly, cooling the elephant down? Can children suggest how a changing environment may affect animals?

Key information:

Children may wish to choose their own favourite animal to study; if this is the case then make sure that the animal demonstrates accessible physical and behavioural traits. Some adaptations that make an organism suited to its environment are very subtle and children may find them difficult to explain.

EVERYTHING CHANGES

 ## LESSON 7: WHY DO LIVING THINGS BECOME EXTINCT?

LESSON SUMMARY

In this lesson children apply their knowledge of how changes in an environment can cause living things to become extinct. By the end of this lesson children know that extinction occurs when the degree or type of changes in the environment are too great for the animals or plants that live there to adapt and survive.

Key vocabulary:

environment, natural selection, adaptation, extinction, population

Resources:

Sheets of A3 paper, secondary sources of information, including the internet, for further research

National curriculum links:

Identify how animals and plants are adapted to suit their environment in different ways and that adaptation may lead to evolution

Working scientifically links:

Identifying scientific evidence that has been used to support or refute ideas or arguments

Learning intention:

To evaluate variables that contribute to the extinction of living things

Scientific enquiry type:

Finding things out using a wide range of secondary sources of information

Success criteria:

- I can use examples to describe how some living things became extinct.
- I can use examples to explain why some living things became extinct.

EXPLORE:

Show children the Extinct and living animals slideshow (Slideshow 1) and ask them to work individually to put the animals and plants into two groups: those that still exist on Earth and those that no longer exist on Earth. Use the word 'extinct' to define a type of living thing that no longer exists. It is important that children can make the distinction between the extinction of types of living things, which means that there are no more individuals of this type alive anywhere, and the death of individual living things.

Ask: *Why do you think that some of these living things no longer exist?*

ENQUIRE:

Explain to children that they are going to choose one of the living things that no longer exists and find out more about it: where and when it lived, and what happened to it. Explain to them that they are going to their work with each other.

The challenges are designed to allow children working in pairs to research an extinct animal, its environment and how it became extinct. The challenges are differentiated by the level of detail that children collect and the depth of analysis of that data. Challenge 1 requires children to connect the characteristics of the animal with its environment. In Challenge 2 children are asked to describe how the animal became extinct, with reference to changes in the environment that it faced. Challenge 3 requires children to explain the process of extinction with reference to adaptation, struggle to survive and the timescales involved.

Key information:

If the internet is not available, children should have access to library books that include information on the dodo, the mammoth and the sabre-tooth tiger.

Challenge 1: Children use the internet to research a living thing that is extinct and create a poster about it, where it lived and how it was adapted to its environment

Ask the children in their pairs to choose one of the extinct animals or plants from Slideshow 1 and to find out what they can about it using secondary sources of information, including the internet. Ask the children to include information about the environment their plant or animal lived in and how it was adapted to that environment.

Ask: *What was the environment like? How was the living thing adapted to its environment? If it was an animal: what did the animal eat? Was it a predator or prey, or both?*

Challenge 2: Children use the internet to research a living thing of their choice that no longer exists, and create a poster about its adaptations to its environment and changes that might have contributed to its extinction

Ask the children, in their pairs, to choose an animal or plant that is now extinct and to find out as much as they can about it using secondary sources of information, including the internet. Ask them to include details about the living thing, including a description of its environment, how it was adapted to that environment, how the environment around it changed and how this change might have contributed to the animal or plant's extinction.

Ask: *How was the animal/plant adapted to its environment? What was it that changed that led to the extinction of this living thing? Was this change related to the environment? If so, which part of the environment?*

Challenge 3: Children use the internet to research a living thing of their choice that no longer exists and create a poster that includes details about why it may have become extinct

Ask the children, in their pairs, to choose an animal or plant that is now extinct and to find out as much as they can about it using secondary sources of information, including the internet. Ask the children to include as much information about the animal/plant, its environment and its adaptations as possible, including an explanation of the changes that caused it to become extinct and why these changes meant that the living thing could no longer survive.

Ask: *What was the change/s in the environment that the living thing had to cope with? Was this something completely new for them to cope with or was it something that they were used to, but that was getting much worse? How could the changes in the environment wipe out the living thing? How might it have avoided becoming extinct?*

REFLECT AND REVIEW:

Provide the opportunity for pairs of children to swap posters with another pair so that they can evaluate each other's work. They can use the Poster feedback table (Resource sheet 1) to help them in their evaluation.

Ask the class to think of the living things that no longer exist and to consider the question: What might the living thing have done to avoid becoming extinct? Explain to them that this is something that they are going to look at in the next lesson.

EVIDENCE OF LEARNING:

Discuss with children the process of extinction, as they make their posters. Can children completing Challenge 1 describe the details of the living thing's environment and suggest how the animal or plant was adapted to survive? Can children completing Challenge 2 describe how the living thing had become extinct and mention the environmental pressures that it faced? Are children completing Challenge 3 able to explain how the living thing became extinct, making direct reference to changes in the environment? Are children able to appreciate that it is the changes in the environment that make it more difficult for living things to survive?

Key information:

The children are only expected to describe the cause, for example, an increase in predators. They do not need to explain why this would lead to extinction.

Key information:

Children should now explain how the cause led to the complete extinction of the animal/plant. Encourage them to refer to the changes in the environment and to explain that not enough of the living things were able to tolerate the new conditions and survive.

EVERYTHING CHANGES

 ## LESSON 8: WHAT DOES IT TAKE TO SURVIVE?

Key vocabulary:

adaptation, environment, evolution, natural selection, inheritance, variation, characteristic, population

Resources:

Model-building materials, if available

LESSON SUMMARY

In this lesson children explore further what living things need in order to survive by designing imaginary animals that have adapted to suit a specific environment. It provides a good opportunity for children to reflect on their learning so far in this module about variation and adaptation. By the end of this lesson children are familiar with the concept of adaptations in living things and recognise that individual adaptations may lead to evolution over a long period of time.

National curriculum links:

Identify how animals and plants are adapted to suit their environment and that adaptation may lead to evolution

Working scientifically links:

Reporting and presenting findings from enquiries, including conclusions, causal relationships and explanations of and degree of trust in results, in oral and written forms such as displays and other presentations

Learning intention:

To describe animal and plant adaptations and explain how the characteristics of the individuals in populations can change over time

Success criteria:

- I can describe adaptations in plants and animals.
- I can make connections between adaptations in plants and animals, and the features of the environment in which they live.
- I can explain how individual adaptations may lead to evolution of new organisms.

EXPLORE:

This Explore activity is designed to link together everything that children have learned in the module so far. The aim is to emphasise that the characteristics that help animals to survive often show variation. In the example here, children should understand that all male deer have antlers to fight and survive, but some individuals have bigger antlers than others.

Display the Deer stag photo (Slideshow 1), which shows a herd of male deer (stags) with different sized antlers.

Ask: *Which characteristics do you think help the deer to survive? How do you think the characteristics help the deer to survive? Which characteristics show variation in this population (that is, which physical features are different between the deer)? Is there any variation in the characteristics that might help the deer to survive?*

Reintroduce children to the term 'adaptation', which refers to characteristics that help a living thing to survive. Remind them that individual organisms that are best adapted to their environment are most likely to survive and so pass their characteristics to their offspring. (Children might need to be reminded of what they learned about selective breeding in Lesson 2 and how selective breeding changes the characteristics of dogs and apples.)

ENQUIRE:

Explain to children that their challenge is to design an imaginary animal that has clearly adapted to survive in its surroundings. The animal can be drawn, as suggested below. Alternatively, depending on available resources, children could make a model. If they are making models, they need to devise a way of labelling the adaptations, for example, by attaching a length of string to the model and linking to a label on a sheet of paper the model is standing on. Either allocate to children or let them choose from one of the Imaginary environments (Slideshow 2) that their animal has to survive in.

The challenges are differentiated by the level of detail and application of ideas required. Challenge 1 asks children to describe the animal and its adaptations. In Challenge 2 children are asked to make explicit connections between the adaptations of the animal and the environment. Challenge 3 requires them to develop this in more depth and to consider further adaptation/s that might occur if the environment changed. Children could work in pairs or in small groups of three or four.

Challenge 1: Children draw a labelled diagram of the animal they have invented to suit the environment

Explain to the children that they should provide as much detail as they can on their diagram.

Ask: *In what ways is your animal adapted to its environment to survive? What adaptations does your animal have to help it survive?*

Challenge 2: Children draw a labelled diagram of the animal that they have invented to suit an environment and explain how its adaptations would help it to survive

Ask the children to put as much detail as they can into their drawings, including information that explains how the animal's adaptations would help it to survive in its environment.

Ask: *How do your animal's adaptations allow it to survive in the environment where it lives? If some individuals of the same type of animal came into the environment but didn't have one of these adaptions, do you think they would survive? Why? If some individuals of the same type of animal came into the environment but had even better adaptations for that environment, do you think they would survive? Why?*

Challenge 3: Children draw a labelled diagram of the animal they have invented to suit the environment and include as much detail as possible about its characteristics and why the animal has those characteristics

Ask the children to include as much detail in their drawings as possible, including describing the characteristics of the animal and why the animal has developed those characteristics to help it survive.

Ask: *What do you think might happen to a population of these animals if the environment changed rapidly, for example, the land suddenly became flooded permanently? What do you think might happen to a population of these animals if the environment changed gradually over a very long period of time, for example, if the area became a desert due to years of very low rainfall?*

Give the children a few minutes to discuss what they think about the two situations and what might happen to animals that live in environments that either suddenly change or that change gradually over time.

Ask: *What do you think the animals would do to survive? Which ones are most likely to survive? Which animals are most likely to have offspring? Why does it matter if the changes happen quickly or slowly? What kind of characteristics might your population develop over a long period of time?*

It is likely that most children will suggest that if the change is rapid then the population will die, but it is possible that variation within the population may allow a small number to survive. These animals would be able to reproduce and have young, so the population doesn't become extinct.

REFLECT AND REVIEW:

Linking in to Challenge 3, start a discussion to reinforce the idea that differences between individuals within a population allow some to survive and reproduce because they are best suited to the new environment.

Ask: *Why is variation important within a population? What could happen to a population if there were no variation and every individual was identical?*

EVIDENCE OF LEARNING:

Question children as they are preparing their imaginary animals and discuss their comments in response to questions about changes in environment. Can children completing Challenge 1 design an animal that was well adapted to its environment? Can children completing Challenge 2 explain their decisions when designing their animal? Are children completing Challenge 3 able to describe the effects on a population of environmental changes happening over time? Are children able to explain why variation is important in a population?

EVERYTHING CHANGES

LESSON 9: WHAT EVIDENCE IS THERE THAT LIVING THINGS HAVE CHANGED OVER TIME?

LESSON SUMMARY

In this lesson children build on their knowledge of fossils from Year 3. They use fossils to examine how plants and animals may have looked in the past and, based on their features, suggest the environment in which they may have lived. By the end of this lesson children will know that fossils provide evidence of living things that no longer exist and information about how the environment has changed over very long periods of time.

Key vocabulary:

fossils, characteristics, environment, adaptation, evolution

Resources:

Collection of fossils, including a fish fossil if possible, or a selection of photographs of fossils, access to secondary sources of information, including the internet, for further research

National curriculum links:

Recognise that living things have changed over time and that fossils provide information about living things that inhabited the Earth millions of years ago

Working scientifically links:

Identifying scientific evidence that has been used to support or refute ideas or arguments

Learning intention:

To recognise that fossils allow us to study things that have lived in the past and provide evidence of evolution

Scientific enquiry type:

Finding things out using a wide range of secondary sources of information

Success criteria:

- I can use fossils to investigate living things that no longer exist.
- I can use my findings to suggest the environment in which those living things must have lived.
- I can describe how looking at changes in fossils can provide evidence of evolution over very long time periods.

EXPLORE:

Ask: *What is a fossil?*

Organise children into groups of three or four and provide each group with the Fossil formation cards (Resource sheet 1). Ask them to match the captions to the images and to sort them into the right order.

Key information:

If possible, use a real fish fossil here. These are widely available and relatively inexpensive.

Display the Fossilised fish photo (Slideshow 1) and ask children to draw a picture of what they think this animal looked like when it was alive. Remind children that fossils are a way of 'looking into the past' at living things that existed millions of years ago, most of which no longer exist.

Review children's drawings of and answers about the fossilised fish, and ask them for suggestions about what type of environment this animal may have lived in.

Ask: *How do we know this animal did not live on land? How much evidence do we have that this animal lived in a watery environment?*

Show children pictures of other Fossilised plants and animals (Slideshow 2).

Ask: *What do these fossils look like? What kinds of environments do you think they lived in? What evidence do you get from looking at the fossil? What other evidence might you need to be more certain?*

ENQUIRE:

Establish that fossils provide evidence of organisms that lived in the past, the types of environments that they lived in and how living things have changed over millions of years. Explain that in the challenges they are going to look at fossils from different times and prepare a presentation to describe what life was like millions of years ago.

The challenges are differentiated according to the level of interpretation children are asked to give through the guided questions and the length of their presentation.

Note:

It is recommended that you print Resource sheet 2 out in colour from Collins Connect.

Challenge 1: Children use secondary sources of information to find out more about a fossil, suggest when and where it may have lived, and prepare a powerpoint presentation

Provide pairs of children with a named photograph of a trilobite (the first picture from Resource sheet 2).

Prompt them to think about the animal, its possible environment and when it lived on Earth.

Ask: *When was this animal living on the Earth? What other types of living things were also on the Earth at the same time? What type of environment did it live in? Why did you give the answers you gave?*

Encourage the children to use secondary sources of information, including the internet, to find out how scientists might answer the questions.

Ask: *How different were your answers from those of the scientists? Were you surprised? Why do you think scientists know so much about these animals?*

Ask the children to prepare a powerpoint presentation of 2–3 slides about their fossil.

Challenge 2: Children use secondary sources of information to find out more about a fossil, suggest when and where it may have lived and how it evolved, and prepare a powerpoint presentation

Provide pairs of children with a named photograph of an ammonite (from Resource sheet 2).

Ask: *When was this animal living on the Earth? What other types of living thing were also on the Earth at the same time? What type of environment did it live in? Why did you give the answers you gave? How many years do you think ammonites lived on Earth? Do you think these animals stayed the same all the time or do you think they changed in some way? Are there any animals living today that are related to ammonites?*

Encourage the children to use secondary sources of information, including the internet, to find out how scientists might answer the questions.

Ask: *How different were your answers from those of the scientists? Were you surprised? Why do you think scientists know so much about these animals?*

Ask the children to prepare a powerpoint presentation of 3–5 slides about their fossil.

Challenge 3: Children use secondary sources of information to find out more about a fossil, suggest when and where it may have lived and how it evolved, and prepare a powerpoint presentation

Provide pairs of children with a picture of Archaeopteryx (from Resource sheet 2).

Ask: *When was this animal living on the Earth? What other types of living thing were also on the Earth at the same time? What type of environment did it live in? Why did you give the answers you gave? How many years do you think Archaeopteryx lived on Earth? Do you think these animals stayed the same all the time or do you think they changed in some way? Are there any animals living today that are related to Archaeopteryx? What seems to be different about this living thing, compared with the other fossils that you have looked at?*

Encourage the children to use secondary sources of information, including the internet, to find out how scientists might answer the questions.

Ask: *How different were your answers from those of the scientists? Were you surprised? Why do you think scientists know so much about these animals?*

Ask the children to prepare a powerpoint presentation of 5–7 slides about their fossil.

REFLECT AND REVIEW:

Ask some of the groups to show their powerpoint presentations.

Emphasise the extremely long timescales that children have been looking at, which makes it difficult to get a good picture of exactly how living things have changed. For things that lived recently it is possible to see more detail of changes.

Display the slideshow depicting Fossil evidence over time (Slideshow 3) and explain that there is a good fossil record of animals that have changed over many years. Explain that evolution describes how animals change over many years due to natural selection and that this natural process of adaptation, over perhaps hundreds of thousands of years, is very similar to the way in which breeds of dogs have been developed over much shorter time frames through selective breeding.

EVIDENCE OF LEARNING:

Can children recognise that fossilisation takes place over millions of years and that it essentially provides an 'image' of a living thing? Do they recognise that fossils provide information about the key physical characteristics of living things? Can children recognise that the characteristics shown by fossils can help us make suggestions about the environment in which the animal lived? Do they note that different fossils may provide evidence for the existence of the same environment (for example, fish fossils and other water-living animals and plants all suggest a body of water)? Do they appreciate that looking at fossils over a period of time can show how an animal has changed and so provide evidence for evolution?

EVERYTHING CHANGES

LESSON 10: HOW DOES NATURAL SELECTION WORK?

LESSON SUMMARY

In this lesson children explore how natural selection works. By the end of this lesson children are able to recognise that natural selection helps to explain how living things have evolved over long periods of time and that these change have led to the organisms that exist today.

Key vocabulary:

environment, adaptation, variation, survival, breeding, generation, population, natural selection, evolution

Resources:

Large pieces of paper, plastic cups, rice, tweezers, tongs, plastic forks, plastic knives, large marbles (if possible)

Health and safety:

Children should be responsible when using the tweezers, forks and knifes, as parts of this type of equipment can be sharp.

Key information:

Make it clear that animals do not change because they want to. It is variation in inherited characteristics that causes changes to living things. Those individuals most suited to their environment are more likely to survive and reproduce, and pass their characteristics to their offspring. Over time the advantageous characteristics become more reliably inherited, so the organism changes and becomes better adapted to the environment.

National curriculum links:

Recognise that living things have changed over time and that fossils provide information about living things that inhabited the Earth millions of years ago

Working scientifically links:

Identifying scientific evidence that has been used to support and refute ideas or arguments

Learning intention:

To describe the process of natural selection

Scientific enquiry type:

Finding things out using a wide range of secondary sources of information

Success criteria:

- I can use the term 'adaptation' correctly when describing living things and the environments they live in.
- I can describe how the process of natural selection works.
- I can explain how living things can change over time due to natural selection and that this change is called evolution.
- I can suggest how natural selection could change existing types of living things over very long periods of time.

EXPLORE:

Display the picture of a Wading bird (Slideshow 1).

Ask: *What are the key features of this bird? What do you think would happen if the bird didn't have these features?*

Organise children into small groups and give each group a copy of the Wading bird graphic organiser (Resource sheet 1). Ask them to fill it in for the wading bird. Some children may need help identifying all of the characteristics; the ones that are most relevant are long legs, the feathers and the beak, but other features can be included.

Collect answers from children about the wading bird's most important features.

Ask: *How did the bird get long legs?*

This question is designed to prompt children's thinking and is answered as the lesson progresses.

Introduce children to the term 'natural selection' and show them the Natural selection animation (Animation 1).

ENQUIRE:

Explain to children that, in small groups, they are all going to play the 'natural selection game'. Explain the rules (Resource sheet 2) so that children are clear what they have to try to do. (Teachers notes are also available; Resource sheet 3). You should control the 'feeding time' within the game and make sure no one is cheating.

All children should play the game and then be given the opportunity to choose a challenge to complete. Challenge 1 is mainly descriptive of the process of the game as a model for natural selection, whereas in Challenge 2 children apply the ideas of natural selection from the game to another animal. Challenge 3 requires that children transfer their knowledge to considering variation in plant characteristics and this challenge should be used as a development and extension activity for those children who clearly understand the process of natural selection in animals, from Challenge 2.

Challenge 1: Children recap on what they have learned about natural selection

After they have played the game, encourage the children to discuss what they have learned about natural selection. Prompt them with some questions.

Ask: *What features in the environment did your feeding animals have to deal with? What happened to the animals that did not have the right characteristics for the job? What happened to the whole population of animals over time?*

Challenge 2: Children apply the ideas of natural selection to another animal

After they have played the natural selection game, ask the children to choose an animal with a distinct characteristic (for example, a giraffe, for its long neck, or a moth, for its camouflage) and describe how features of that animal's environment influenced the development of this characteristic.

Encourage the children to apply a similar step-by-step process to that outlined in the Natural selection animation (Animation 1) and the ideas from the natural selection game. Prompt them by asking some questions.

Ask: *Which animal have you chosen and which characteristic have you focused on? How has natural selection caused this characteristic to develop? Do you think this process happened quickly?*

Challenge 3: Children apply what they have learned about natural selection in animals to plants

Encourage the children to apply the ideas from the natural selection game, and what they know about natural selection in animals, to consider what happens in plants. Ask them to pick a plant that has a particular characteristic and to describe how it may have changed over time.

Ask: *Which plant have you chosen to focus on? Why? How does the plant use its features to survive? How has natural selection caused this feature to develop?*

Key information:

If possible, children should be encouraged to identify a plant and its characteristics on their own. However, the following examples are perhaps most accessible and can be given to children without impacting on the value of the task: the spikes on holly; deep and wide roots in some desert plants; sugary nectar in some flowering plants.

REFLECT AND REVIEW:

Organise children into pairs and give each pair a copy of the Environmental conditions card sort (Resource sheet 4). Ask one child in the pair to select randomly an animal from the collection of animal cards and the second child in the pair to select at random an environment from the collection of environment cards.

Encourage children to discuss the main aspect of the environment that affects the animal and how the animal that has been selected would need to change over many generations to fit the new environment. For example, if the animal were a rabbit and the environment were the desert then the rabbit may, over many generations, lose its fur and develop a method for cooling itself down and for retaining water.

Ask some pairs to share ideas with the rest of the class.

Key information:

It is important to continue to reinforce that natural selection takes place over many, many generations. Children must not think this process happens within one generation.

EVIDENCE OF LEARNING:

Listen to children's descriptions of natural selection as they play the game and discuss with them how their animals are changing over time. Are children able to understand that natural selection takes place over many generations? Can they explain how the game was modelling natural selection? Are children able to apply their knowledge of natural selection to different living things, including plants? Do they use the terms 'natural selection' and 'adaptation' correctly, in context? Are they able to give specific examples of natural selection and adaptation?

EVERYTHING CHANGES

ENRICHMENT LESSON 1: HOW CAN ONE TYPE OF ANIMAL BECOME TWO?

LESSON SUMMARY

In this lesson children apply their knowledge of natural selection to the more complex process of speciation. By the end of this lesson children know that speciation occurs as a result of a geographical barrier and natural selection. This lesson is linked to Lesson 10 and should be introduced after that lesson.

Key vocabulary:

species, offspring, infertile, speciation, geographical barrier, natural selection

Resources:

Access to secondary sources of information, for further research if there is time

National curriculum links:

Identify how animals and plants are adapted to suit their environment in different ways and that adaptation may lead to evolution

Working scientifically links:

Identifying scientific evidence that has been used to support or refute ideas or arguments

Scientific enquiry type:

Finding things out using a wide range of secondary sources of information

Learning intention:

To describe the process of speciation

Success criteria:

- I can define the word 'species'.
- I can define the word 'speciation'.
- I can explain how speciation occurs.

EXPLORE:

Show children a picture of Charles Darwin (slide 1 of Slideshow 1) and ask any of them if they know who he is. Once children (or you, if no one can name him) have identified the picture as Charles Darwin, explain to them that they are going to repeat the experiment that he did to work out something really important about living things.

Display a picture of the Galápagos Islands (slide 2 of Slideshow 1) and explain to them that this was where Darwin did some of his experiments, but that they have to stay in the classroom. Organise children into pairs or group of three and give them the Finches card sort (Resource sheet 1). Ask them to try to match each finch with the food types that it eats. The size of the finches' beaks should be an indication of which food group each one prefers.

ENQUIRE:

Explain to children that the activity they have just done is one of the things Charles Darwin did to help work out the theory of natural selection. Explain that Charles Darwin is the scientist who is most well known for his theory of natural selection (and that, by coincidence, another scientist called Alfred Russell Wallace came up with the idea of natural selection at the same time as Darwin). Explain to them that since Darwin's time many other scientists have studied and are still studying this idea of how living things evolve. (Children may wish to find out more about Darwin and Wallace as a homework activity.)

Continue to explain to children that these different finches really do exist and that they evolved from one finch, through the process of natural selection. Explain that the original finch lived on one large island, which was surrounded by lots of smaller islands. Over many years these islands moved away from each other, due to changes in the Earth's crust, and the birds were no longer able to fly between islands, so they became permanently separated from each other. Each island had its own different food supply and eventually the birds on each island developed a type of beak that was suited to collecting the types of foods found on that island.

Establish with children that the type of food supply available is one example of how the environment can influence which individuals survive and so how, over time, animals adapt.

Next, show the class the Speciation animation (Animation 1). Explain that the term 'speciation' refers to one species evolving into two or more different species because of changes in the environment. Stress that this process takes a very long time. Ensure that children understand that in the example shown in the animation, speciation was driven by whether the ground was covered in grass or not and that this was the environmental change that influenced how the beetles adapted.

Prepare the class for the challenges by organising children into small groups and asking them to choose an animal. Explain that they now have to imagine a population of this animal that has split into two groups because of environmental changes, just like the finches and the beetles.

The challenges require children to use the ideas of evolution and speciation to imagine, describe and explain what might happen to their animal. The challenges are differentiated by the level of detail and analysis required as well as the application of scientific ideas and the correct use of scientific vocabulary.

Challenge 1: Children produce a cartoon strip to describe how a population of animals could be separated into two

Ask the children to draw a cartoon about how a population of the animal they have chosen might separate into two different species. The cartoon should include information on how the animals were separated, what both groups might look like and some information about how the environments of the two groups are different.

Ask: *What is a geographical barrier? Which geographical barrier have you chosen for your cartoon strip? How is the environment different on both sides of your geographical barrier?*

Challenge 2: Children produce a cartoon strip showing how speciation could take place with the animal of their choice and include the terms 'natural selection', 'adaptation', 'evolution' and 'geographical barrier'

Ask the children to draw a cartoon about how a population of the animal they have chosen might separate into two different species. The cartoon should include information on how the animals were separated, what both groups might look like and some information about how the environments of the two groups are different. Their cartoon should include the terms 'natural selection', 'adaptation', 'evolution' and 'geographical barrier'.

Ask: *What are the changes in the environment on this side of the geographical barrier? How have your animal's physical features changed to enable its survival in this environment? Would this be a fast process or a slow process?*

Challenge 3: Children produce a cartoon strip that tells the story of a scientist discovering two species of animal and working out that the two animals have evolved from a single species by speciation

Ask the children to draw a cartoon that charts the work of a scientist who has discovered two species of animal and worked out that the two populations have evolved from a single species, by the process of speciation.

Explain to the children that their cartoon should include details about the scientist's observations of the animals and their environments, as well as explanations of why the scientist thinks speciation happened and how they could determine if the two populations really were different species.

Ask: *What are the observations that the scientist has made and how do they help to explain what the scientist thinks has happened? If we breed two animals of the same species what happens? If we breed two animals of a different species what happens? How could we find out if a geographical barrier has caused the evolution of two new species?*

RELECT AND REVIEW:

To summarise the lesson, show the class pairs of photographs of Closely related animals (Slideshow 2), for example, polar bear and grizzly bear, and ask them to suggest how these two different species may have developed from a single species.

As part of the discussion, highlight the main points of speciation: that there must have been a geographical barrier; that there must be at least two different environments; and that natural selection must have led to the development of two different species. Explain to children that speciation is responsible for the wide range and variety of species of plants and animals on the Earth.

EVIDENCE OF LEARNING:

Discuss the process of speciation as children carry out the challenges. Can children describe a geographical barrier and suggest different environmental conditions on each side? Can they describe how natural selection could have taken place differently on each side of the barrier? Are children able to explain how we could demonstrate that the two different populations are different species? Are they able to use the terms 'speciation', 'natural selection', 'adaptation', 'evolution' and 'geographical barrier' correctly, and identify examples in their own comic strips?

Key information:

The key message in this task is that for speciation to occur there must be a geographical barrier separating the population and that conditions must be different on both sides of the barrier.

Key information:

Two groups of animals are classified as different species if breeding one of each group together results in offspring that are infertile, that is, the offspring cannot breed and produce their own young. For example, two ligers (a cross between a lion and a tiger) cannot breed to produce more ligers.

DANGER! LOW VOLTAGE

Health and safety:

Zinc-carbon and zinc-chloride batteries are the most suitable for this type of activity. Alkaline and rechargeable batteries may become hot enough to cause burns if there is a short circuit. Follow guidance in Be Safe! section 14.

INTRODUCTION

In this module children develop their understanding of electrical circuits and build on the work in the Year 4 module. They construct circuits with an increasing number of components and contrast the effects this has on the function of the components. They role play the flow of electricity through a basic circuit and one that includes fuse wire, to model the effect that this has on other components.

The children learn to use the recognised electrical symbols to record circuits, particularly as the circuits become more complex. They research how electricity is generated both traditionally using coal and gas, and by renewable resources, and investigate how electricity is transmitted across the country, and what sort of electricity generating plant they might site in their locality.

In the extension lessons children apply their knowledge to construct circuits for real life contexts, and then report on and present how they did this, as scientists, to the class and to a third party .

When working scientifically, children carry out illustrative practicals, describe circuits using scientific language and record them using the recognised symbols. In lessons 5 and 6 children use secondary sources of information to answer questions about how mains electricity is generated.

Enrichment Lessons 1 and 2 should be taken as two blocks of time together, as should Enrichment Lessons 3 and 4. Although there are specific things to complete in each lesson, extra time might be needed to complete the challenges in the second of these two pairs lessons. As part of a science display a space should be set aside as a question wall for children to post their questions during the module.

Key vocabulary:

cell, battery, lamp, wire, buzzer, motor, circuit, current, filament, electrical insulator, electrical conductor, mains electricity, terminal, switch, toggle switch, push switch, slide switch, tilt switch, trembler switch, pressure switch, reed switch, series circuit, resistance, resistor, current, circuit diagram, recognised symbols, generate, generator, coal, gas, oil, fossil fuels, nuclear, biomass-fired power stations, wind turbine, wave hub, tidal flow, hydro-electric, grid, pylon, transmission, transformer, solar panels

FACT FILE:

See also the module introduction for Year 4, Module 3, Switched On.

Electricity is a flow of electrons and this flow produces an electric current. A single battery is called a cell. Batteries are formed when a number of cells are grouped together. Once all the chemicals in a cell have reacted together, then no more extra electrons can be produced and the cell is 'dead'. A chemical reaction in a cell results in extra electrons at the negative terminal and a shortage at the positive terminal. As electrons are negatively charged they are attracted to the positive terminal when the two terminals of a cell are connected in some way. The current in all parts of a circuit is instantaneous and equal. Electrons keep flowing through the circuit and they are not used up in the creating of light, movement or heat.

The contents of a cell vary depending on the type of cell.

Cells, switches, lamps, buzzers, motors, etc., are called components of circuits. They have two connection points called terminals. Both terminals must be connected to a power source's terminals in a loop for the electrons to flow and for the circuit to be complete. It is possible to construct a circuit and light a lamp using only one piece of wire, a cell and a lamp.

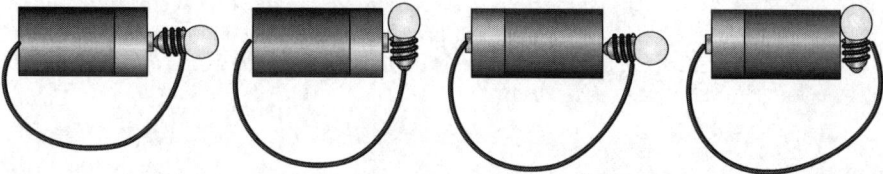

There are four more ways to complete the circuit by reversing the wire in the above diagrams.

Voltage is the driving force that causes current to flow around a circuit – 'the push'. It is the difference in potential energy between the positive and the negative terminals. As the voltage increases so does the work the current can do. For example, the higher the voltage in a circuit the brighter a lamp shines and the faster a motor turns. Matching the voltage of cells and components in circuits is essential to avoid 'blown' lamps and 'burned out' motors. Voltage is measured in volts.

Resistance is the measure of the difficulty electrons have in flowing through a material. It is much easier for the electrons to flow through a thick wire than through a thin wire. The resistance is much higher in a thin wire. Lamp filaments are very thin wire and have a high resistance. This converts the electrical energy into heat energy and results in the filament glowing white hot, therefore giving light as well as heat. Resistance is measured in ohms.

These symbols are the recognised scientific conventions for representing circuits:

Cell		Three-way switch	
Battery (2 or more cells)		Push switch	
Lamp		Resistor	
Buzzer		Variable resistor	
Motor		Wires	
Open switch		Where wires join	
Closed switch		Where wires cross	
Two-way switch			

The scientific convention for circuit diagrams always shows wires as straight lines with right angle turns.

Mains electricity is generated at power stations at very high voltages, although power station outputs are usually measured in megawatts (1 MW = 1 million watts). Watts are the power in a circuit. For example, a 100 watt lamp will be much brighter than a 60 watt lamp. Electricity is transmitted to houses, offices and factories through a series of large cables either suspended from pylons or laid underground. This system is called the national grid. As the electricity gets closer to homes it is systematically stepped down by sub-stations containing transformers that gradually reduce the voltage from the hundreds of thousands of volts generated in power stations to the 230V used in the home. Mains electricity is a big circuit so when a kettle is plugged in at home, the circuit is complete from the house to the power station and back again. Power stations are like a large cell. Electricity is generated by:

- burning fossil fuels such as coal, gas and oil
- biomass energy – burning wood chip and domestic and commercial waste
- nuclear reaction
- movement of water – tidal flow, wave motion, falling water
- solar energy
- wind energy

COMMON MISCONCEPTIONS:

See also the module introduction for Year 4, Module 3, Switched On.

Children think that there is only a flow of electrons in the wire in one direction to the lamp and that the return wire is unnecessary or 'empty' because the electrons have been 'used up' lighting the lamp.

Children often think that a switch has to be between, and close to, the cell and the lamp for it to light a lamp.

DANGER! LOW VOLTAGE

LESSON 1: HOW MANY SIMPLE CIRCUITS CAN YOU MAKE?

Key vocabulary:

cell, battery, lamp, wire, buzzer, motor, circuit, current, filament, electrical insulator, electrical conductor, mains electricity, switch, terminal, electrons

Resources:

Commercially produced energy stick or human circuit ball (available from primary science equipment suppliers, for example, TTS), 1.5V cells, lamps, lamp holders wire, thin tinfoil strips, cell holders, (one of each component between two), magnifiers, digital magnifier, modelling clay (which is useful to anchor a cell while the circuits are constructed), extra wire, small screwdrivers, mini whiteboards

Health and safety:

Zinc-carbon and zinc-chloride batteries are the most suitable for this type of activity. Alkaline and rechargeable batteries may become hot enough to cause burns if there is a short circuit. Do not use button batteries. Follow guidance in Be Safe! section 14. Although there are no specific control measures that need to be taken by children in this lesson, use the opportunity to remind them that batteries are a safe source of electricity; they should not experiment using mains electricity. Remind children of the dangers of mains electricity.

LESSON SUMMARY

In this lesson children revise and build on their work from Year 4 on how to construct simple circuits. By the end of the lesson they understand and will have demonstrated in an illustrative practical how electricity flows through a complete circuit.

Preparation required:

As part of a science display for this module, set up a section for children to use as a question wall, where they can write the questions that they may have about electricity on sticky notes. These questions can then be answered during the course of the module.

National curriculum links:

Use recognised symbols when representing a simple circuit in a diagram

Learning intention:

To represent a simple circuit in a diagram and describe how it works

Scientific enquiry type:

Carrying out simple comparative and fair tests

Working scientifically links:

Recording data and results of increasing complexity using scientific diagrams and labels, classification keys, tables, scatter graphs, bar and line graphs

Success criteria:

- I can construct a simple circuit.
- I can represent my circuit in a labelled drawing using the correct scientific language.
- I can represent my circuit in a circuit diagram using the recognised symbols.
- I can explain how the terminals are important when constructing a circuit to light a lamp.

EXPLORE:

Ask children, in pairs, to cut out the Ten true or false statements (Resource sheet 1) about electricity and circuits, to sort them according to whether they think they are true or false, and to select the three most important true statements. Use a commercially produced energy stick or human circuit ball to create a circuit with the whole class linking hands.

Ask: *What has to happen to make the stick/ball work? What happens when the link is broken in one of several different places? (Not at the same time.)*

Children need to know that a complete circuit with no breaks is needed to make the components work, but it is not necessary at this stage for children to be able to explain why. Ask them if they now think they should change any of their true/false statements or if they should select different true statements as the most important. Keep these for the Reflect and review part of the lesson.

ENQUIRE:

Tell children that their challenge is to make simple, complete circuits to light a lamp. Working in pairs, they are going to be given a cell, a lamp, a piece of tinfoil and some modelling clay. Explain that the modelling clay is to hold the cell firmly on the table. The challenges are differentiated by the number of different circuits they need to make and record.

Ask children why they think they have been given a strip of metal foil rather than a piece of wire.

Encourage children to think about how they can construct a circuit with the components and resources that they have been given, what they have to remember about the positions of the contacts on each component and how to connect these contacts to make a complete circuit with only one strip of metal foil.

Ask children to use a magnifier to look at a lamp, then to draw what they can see inside the glass globe. When they have lit the lamp ask them to draw their circuit on a mini whiteboard.

Key information:

Children may think that there is only a flow of electrons in the wire in one direction to the lamp and that the return wire is unnecessary or 'empty' because the electrons have been 'used up' lighting the lamp.

The challenges are presented on the Challenge slides to be displayed on the board, or printed out and placed in the centre of the table.

Challenge 1: Children construct two simple circuits

Give the children time to discover two other ways to complete the circuit and draw what they have done.

Challenge 2: Children construct four simple circuits

Give the children time to construct four different circuits that will light the lamp, and to record these in labelled diagrams.

Challenge 3: Children construct many simple circuits

Give the children time to construct as many circuits as they can that will light the lamp, and to record these in labelled diagrams. Encourage the children to explain why they think the lamp lights.

After they have created their circuits, ask the children what extra components would make the task easier and give them extra wires with crocodile clips, lamp holders and cell holders. Ask the children to look for the two separate terminals or connections on these components so that they understand the need to maintain the circuit to ensure the flow of electrons. Give the children some time to remake one of their circuits using these components.

Key information:

Metal foil is used instead of plastic coated wire so that children understand that it is the metal that is conducting the electricity and that the plastic coating is there as an insulating material, particularly as the voltages increase.

Introduce the electrical symbols for a cell, lamp and wires (Circuit diagram symbols; Slideshow 1) and ask all the children to draw their circuit again. Those completing Challenge 1 can use the Circuit diagram symbols drag and drop interactive (Interactive 1) on the interactive whiteboard to recreate their circuits.

Key information:

Please refer to Simple circuits (Resource sheet 2) for diagrams of the circuits and a lamp.

REFLECT AND REVIEW:

Enourage children to discuss the lesson and their challenges with their partners, and to write two true/false statements about terminals.

Ask: *Do you want to review any of the three true statements from the beginning of the lesson? What have you learned about the terminals on components? Why do you think it might be better to use electrical symbols rather than labelled diagrams to represent circuits?*

Explain that the symbols used to represent electrical components are internationally agreed and are recognised as a science convention worldwide.

EVIDENCE OF LEARNING:

Listen to children as they sort the true/false questions (Resource sheet 1). Observe them as they build and then draw their circuits.

Can they construct simple circuits? Do they understand that components need two terminals and how they must be connected to complete circuits? Are children confident in constructing circuits that demonstrate the flow of electricity (electrons)? Can they represent their simple circuits using the recognised electrical symbols?

DANGER! LOW VOLTAGE

LESSON 2: WHAT DOES A SWITCH DO?

Key vocabulary:

cell, battery, lamp, wire, buzzer, motor, circuit, current, electrical insulator, electrical conductor, mains electricity, terminal, types of switches including toggle, push, slide, tilt, plunger, trembler, pressure

Resources:

A2 paper, 1.5V cells, lamps, wire, crocodile clips, toggle switches, slide switches, push switches, lamp holders, cell holders, small screwdrivers, wire strippers, match boxes, metal foil, paper fasteners, paper clips, film canisters, small ball bearings, card, adhesive tape or glue, hand drills, drill bit, examples of mains switches

Health and safety:

Mains electrical components used as examples of switches should not be connected to the mains supply. Small bedside-type lamps should only be operated by a member of staff. Follow guidance in Be Safe! section 12 'Making things' for the use of hand drills.

Key information:

Children often think that a switch has to be between the cell and the lamp in a circuit for the lamp to light.

LESSON SUMMARY

In this lesson children make and control simple circuits using purchased switches and classroom-made switches. By the end of the lesson they will have demonstrated how to use different types of switches to control circuits.

Preparation required:

Sufficient electrical components for sharing in pairs. This lesson requires a variety of switches for the circuits. Children make as many examples as possible of the switches that they encounter in everyday life, for example, room light switch, wall socket with switches, table lamp push switch built into the lamp holder, button slide switch built into the wire of a lamp, push switches on radios, computers. (See safety note below.)

National curriculum links:

Compare the functions of different components, giving reasons for variations in how components function, including the brightness of bulbs, the loudness of buzzers and the on/off positions of switches, and use recognised symbols when representing a simple circuit in a diagram

Learning intention:

To use a switch in a simple circuit, show it in a diagram and describe how it works

Scientific enquiry type:

Carrying out simple comparative and fair tests

Working scientifically links:

Recording data and results of increasing complexity using scientific diagrams and labels, classification keys, tables, scatter graphs, bar and line graphs

Success criteria:

- I can control components in a circuit with a switch.
- I can represent my circuit in a circuit diagram using the recognised symbols.
- I can recognise that there are many different types of switches.

EXPLORE:

Ask children in their table groups to make a list, using a blue felt tip pen and an A2 sheet of paper, of all the electrical switches in the classroom. Allow them 1 minute to complete their list, then ask children to list, using a red felt tip, the switches in their kitchens at home (allow 1 minute) and next, using a black felt tip, to list the switches in their bedrooms (again, allow 1 minute).

Ask children if this was difficult to complete in just 1 minute for each room. Show them the images of items that may be found in each room (Resource sheets 1, 2 and 3) and see how many items with switches children identified in their rooms.

Ask: *Where were the most switches? Why do we need switches on electrical items?*

It is necessary for children to know that switches are a way of controlling the flow of electricity (electrons) in a circuit. Switches make or break a circuit, turning the component (for example, lamp) on when the switch is closed and off when the switch is open. Display the Switches diagrams slideshow (slide 1 of Slideshow 1), which shows the symbols for denoting a switch as closed (on) and open (off), and different types of switches, so that children can see how to represent switches on circuit diagrams.

ENQUIRE:

Tell children that they are going to construct circuits using switches in different ways. The challenges are differentiated by the complexity of the circuit and recording expectation. Ask children to use a commercially produced toggle switch to construct a circuit to light a lamp. Show the correct symbol for a switch on the whiteboard.

The challenges are presented on the Challenge slides to be displayed on the board, or printed out and placed in the centre of the table.

Challenge 1: Children investigate the positioning of a switch in a circuit

The children work in threes. Ask the children what happens if the switch is connected in different places within the circuit. Encourage the children to use other components, such as a buzzer and motor, instead of the lamp to see how they work. Display the Motor and buzzer slideshow (slide 2 of Slideshow 1) to show them how to depict a motor and buzzer on a circuit diagram. A small disc of cork with a colour mark will help show the direction of travel of the motor spindle. Ask the children if all the components will still work if their connections are reversed.

Challenge 2: Children investigate the use of two switches and another lamp

The children work in threes. Ask the children to use two switches in their circuits and then to add another lamp. Can they light both lamps together and individually? Would a different type of switch be needed?

Challenge 3: Children investigate the use of a slide switch to control two components

The children work in threes. Ask the children to control two components of their choice with a single three-way slide switch. They should show the circuit diagram on the board.

Once children have finished their challenge, give them the materials to make two of each type of switch. Ask the children to test them in circuits. The instructions are provided on the Pressure switch, Slide switch, Tilt switch, Trembler switch and Peg switch sheets (Resource sheets 4–8, respectively). Keep the switches, which can be used in Enrichment Lesson 1.

REFLECT AND REVIEW:

Give children the mains switches to look at.

Ask: *Where are the connections? Why are certain materials used in the construction?* Ask children, in pairs, to write a text message to another pair of children that explains what switches do and also poses a question that they want to ask about switches. Add questions to question wall.

EVIDENCE OF LEARNING:

Observe children as they construct their circuits. Read their text messages.

Do children understand what the function of a switch is in a circuit? Can they represent increasingly complex circuits using the recognised symbols?

DANGER! LOW VOLTAGE

 ## LESSON 3: HOW STRONG IS YOUR RESISTANCE?

LESSON SUMMARY

In this lesson children add different components to electrical circuits and role play the flow of electrons in a circuit to explain the idea of resistance. By the end of this lesson they understand that the number of components in a circuit affects the way the circuit performs.

Preparation required:

Sufficient electrical components for children working in pairs to make circuits with several cells, lamps, buzzers, motors and wires. Graphite pencil leads or propelling pencil leads mounted on lollipop sticks, resistance wire and/or different thicknesses of fuse wire.

Key vocabulary:

cell, battery, lamp, wire, buzzer, motor, circuit, series circuit, switch, resistance, resistor, electrical insulator, electrical conductor, mains electricity, terminal, current

Resources:

1.5 V and 4.5 V cells, lamps, wire, crocodile clips, switches, lamp holders, cell holders, small screwdrivers, wire strippers, pencil or propelling pencil leads stuck to lollipop sticks, resistance wire and/or different thicknesses of fuse wire

Key information:

Resistance is the measure of the difficulty that electrons have in flowing through a material.

Voltage is the 'push' that causes current to flow round a circuit.

National curriculum links:

Associate the brightness of a lamp or the volume of a buzzer with the number and voltage of cells used in the circuit, compare and give reasons for variations in how components function, including the brightness of bulbs, the loudness of buzzers and the on/off position of switches, and use recognised symbols when representing a simple circuit in a diagram

Learning intention:

To demonstrate the effects of changing the current flowing through components in a circuit

Scientific enquiry type:

Carrying out simple comparative and fair tests

Working scientifically links:

Reporting and presenting findings from enquiries, including conclusions, causal relationships and explanations of and degree of trust in results, in oral and written forms such as displays and other presentations

Success criteria:

- I can describe, using correct scientific language, how changing the number and types of components in a circuit affects how they operate

- I can give reasons, using correct scientific language, for what happens to lamps, buzzers and motors when cells are added to a circuit.

- I can explain, using correct scientific language, what happens to lamps, buzzers and motors when a resistor changes the flow of electricity in a circuit.

EXPLORE:

Share some of the text messages about switches that children wrote in Lesson 2. Ask children what they think the difference is between the switches they used in the last lesson, a classroom light switch and a switch that turns off all the lights in the school. Why wouldn't a single cell make a classroom light work?

It is necessary at this stage to explain that electrical components are rated in several ways and that one of these is voltage, which is measured in volts. Ask children if they know what the mains voltage is in the UK (it is 230 V) and what voltage they have been working with when using a cell in their circuits. Re-emphasise the dangers associated with the misuse of mains electricity and that the cells children are using are a perfectly safe way to investigate electricity. Ask children to look at their components to find if they are labelled with a voltage and what that voltage is.

Key information:

So far children have been using individual cylindrical 1.5 V cells. Cells of 4.5 V and 9 V are also safe to use but must be matched to the voltage of other components in a circuit. Lamps, buzzers and motors will operate with a single cell. Lamps are usually 1.5 V, motors are often rated at up to 4.5 V and buzzers up to 6 V, but check before using.

Ask: *What do you think will happen if two lamps are wired into your circuit with a single cell? What about two cells with one lamp? How many volts will there be with two cells? What will happen to the lamp?*

ENQUIRE:

Explain to children that they are going to investigate the effect of adding components to a circuit, remembering to match voltages of cells and other components. The challenges are differentiated by the complexity of the task and the level of explanation expected, moving from simple description to explanation.

The challenges are presented on the Challenge slides to be displayed on the board, or printed out and placed in the centre of the table.

Challenge 1: Children investigate increasing the number of components in a circuit by constructing a circuit with a single 1.5V cell, switch and buzzer, and then increasing the number of buzzers and recording what happens

The children work in threes. Give the children some time to construct a circuit with a buzzer. Ask them to find out what happens if they add a second 1.5V cell to this circuit and then a third. (Check the buzzers are rated up to at least 4.5V.) When the children have completed their task and made a note of what happens, ask them to describe what they observed in an "If...then..." sentence.

Challenge 2: Children investigate the effects on components in a circuit of using higher voltage cells that match the acceptable voltage of the components

Ask the children what they think would happen if they used a 4.5V cell with a single lamp and give them time to construct the circuit. Then ask them to investigate what happens when they add extra components to their original matched voltage circuit. Ask them to explain what they find in an "If.... then....because..." sentence.

Challenge 3: Children investigate what happens when very thin wire is used as a resistor in a circuit

Ask the children to construct a circuit using a 1.5V cell, lamp, switch, wire and a length of resistance or fuse wire. Ask them not to complete the circuit but to leave a gap between the resistance/fuse wire and the ordinary wire. Then ask them what happens when they complete the circuit by touching the resistance/fuse wire in different places along its length with the ordinary wire. What happens if they use a different thickness of fuse wire? Ask them to explain this in an "If....then...because..." sentence. Ask what would happen to a motor and buzzer in a circuit to which resistance/fuse wire was added.

REFLECT AND REVIEW:

Ask each challenge group to draw one of the circuits on the board or IWB and to explain what is happening.

Introduce the symbol for a resistor, by displaying the Circuit diagrams symbols slide (Slideshow 1). Ask children about the applications for using a resistor in a circuit.

Explain to children that they are going to role play being electrons, moving first through an ordinary piece of wire in a circuit and then through much thinner wire, to find out what happens. It is necessary to explain to children that electrons are very small particles that carry the electric charge in a circuit. Show children scene 1 of the Flow of electrons video (Video 1).

Split the class into three groups; form two groups into concentric circles with plenty of space between them. Give some children in these circles labels as a lamp, a switch and a cell. The third group of children represents the electrons and they should spread out between the circles.

Show children scene 2 of Video 1. When the switch is closed the electrons move around between the circles and the lamp is lit. The electrons are the flow of electricity or the current. Open the switch to stop the flow and move one small part of each circle closer to each other so there is a much smaller gap for the electrons to flow through. Close the switch and ask children who represent electrons what is happening to them when they reach the narrow part of the circles. Show scene 3 of the video.

EVIDENCE OF LEARNING:

Observe children as they make their circuits. Watch as they carry out the role play.

Can children predict and notice that adding components to a circuit affects how components operate? Can they match the voltage of cells with components? Can they use the role play model to explain why it is more difficult for electrons to move through the fuse wire? Do they use the correct scientific vocabulary? Are they able to draw these circuits in a diagram using the recognised symbols?

Key information:

It may be necessary to have examples of resistors to stimulate children's ideas, for example, the volume control on a radio, a light dimmer switch and remote control handsets that speed up and slow down models.

Key information:

Don't show scene 3 until children who are electrons have tried to pass through the narrower part of the circles.

DANGER! LOW VOLTAGE

 ### LESSON 4: DO YOU KNOW YOUR CIRCUIT DIAGRAMS AND CAN YOU CONSTRUCT WORKING CIRCUITS FROM THEM?

LESSON SUMMARY

In this lesson children consolidate their learning on circuits and recognised electrical symbols from the previous three lessons. By the end of this lesson they are able to use the recognised symbols both to draw and to construct circuits of increasing complexity.

Preparation required:

Sets of laminated cards for the Circuit diagram bingo game (Resource sheet 1) and cut out cards for the Going round in circles game (Resource sheet 4).

Key vocabulary:

1.5V cells, battery, lamp, wire, buzzer, motor, circuit, series circuit, switch, resistance, resistor, electrical insulator, electrical conductor, mains electricity, terminal, current, circuit diagram, recognised symbols

Resources:

1.5V cells, lamps, wire, crocodile clips, switches, lamp holders, cell holders, small screwdrivers, wire strippers

Key information:

Electricity is the flow of electrons (current) around a circuit. In the Explore section children model this before the Enquire challenges.

National curriculum links:

Associate the brightness of a lamp or the volume of a buzzer with the number and voltage of cells used in the circuit, compare and give reasons for variations in how components function, including the brightness of bulbs, the loudness of buzzers and the on/off position of switches, and use recognised symbols when representing a simple circuit in a diagram

Learning intention:

To demonstrate how circuits can be represented in, and constructed from, diagrams

Scientific enquiry type:

Carrying out simple comparative and fair tests

Working scientifically links:

Recording data and results of increasing complexity using scientific diagrams, classification keys, tables, scatter graphs, bar and line graphs

Success criteria:

- I can represent my circuit using the recognised symbols in a diagram.
- I can explain what happens to lamps, buzzers and motors when cells are added to a circuit.

EXPLORE:

Ask the whole class to form a circle and to join hands. (Include yourself if you think it appropriate.) Tell them that the idea here is to model what happens in an electric circuit. One person, either yourself or a child, acts as the cell and squeezes the hand of the next child who passes this squeeze on. Ask a child who is halfway round the circle to buzz when the squeeze reaches her/him and to pass the squeeze on. When it reaches the person who began the squeeze, he/she says stop. This process should be timed and possibly repeated to improve the timing. Ask children why the activity represents what happens in an electric circuit and then ask them why it does not, emphasising the timing. Press a classroom light switch and try to time how long it is before the light comes on.

ENQUIRE:

Tell children that they are going to use what they have learned about circuits and circuit diagrams in some games and quizzes. The challenges are differentiated by the level of the tasks, from children knowing what the symbols represent to children using them to evaluate and plan circuits.

The challenges are presented on the Challenge slides to be displayed on the board, or printed out and placed in the centre of the table.

Challenge 1: Circuit diagram bingo!

Tell the children that they are going to play circuit diagram bingo and then construct the circuits to check that they work. Give each child a bingo card (Resource sheet 1). Then divide the individual symbol cards between the bag and a pile for the children to pick from when a component is called out. The caller uses the master card on which to place the symbols when he/she has called them out.

Choose one child to be the first caller. The caller takes a symbol card out of the bag and says what it is without showing the card to the other children. He/she then places it on the master card in the correct place. If a player thinks that they have that component on their bingo card then they take the correct symbol card from the spare card pile and use it to cover the symbol on their card. When a child thinks that they have covered all the components needed to complete a circuit, they say "connected". They then show their circuit using the cards they have collected and check that they are correct by building the circuit with the actual components. Repeat, with the children taking it in turns to be the caller. Both caller and players have to know what each symbol represents.

Explain to the children that for this game a lamp includes a lamp holder and a cell includes a cell holder. This extends the game because the children need at least two wires to complete a circuit when using a lamp holder.

Circuits using a buzzer with wires attached need only a buzzer and a cell. The children do not have to use all the components they have covered if they can complete a circuit, for example, without another wire card.

A resistor or variable resistor has to be used with either a lamp, motor or buzzer. If it is included in a circuit without one of these it will cause a short circuit.

Challenge 2: Children check a set of circuit diagrams and say whether the circuits will work

The children work in pairs. Give the children a copy of Will it work? (Resource sheet 2), which includes a number of different circuits. Ask the children to look at each and, for those that they think will not work, to write an explanation and correct the diagram. Ask the children to pass their sheet on to a partner to check, then give each pair some time to construct three of the working circuits.

Challenge 3: Children draw circuit diagrams from lists of components

The children work in pairs. Give the children a copy of Make it work (Resource sheet 3). Ask them to draw the circuits using the components listed and to record whether or not each circuit will work and why. The children check another pair's work. When they have completed the sheets, give the children time to construct three of the working circuits.

REFLECT AND REVIEW:

Ask children why they think the recognised symbols are used. Then, ask children, working in small groups, if they can think of three applications for which the symbols make representation easier than drawing a picture.

Split the class into three groups and play the Going round in circles game (Resource sheet 4) to assess their knowledge. Pick a child to read out a question from their cards. The other children look through their cards to find the answer. When they have found the card with the answer on, they read out the next question, which is on the opposite side of the card. The game continues until you return back to the starting question.

EVIDENCE OF LEARNING:

Do children recognise the symbols? Are they able to construct complete circuits using the symbol cards only? Are they able to construct circuits from the symbol cards using real components? Are they able to draw the complex circuits that they construct using the recognised symbols?

Key information:

As children cannot represent the four sides of a circuit diagram with their symbol cards it is acceptable for them to say "connected" when they have two wire cards if they can show the circuit will works using real components.

DANGER! LOW VOLTAGE

 ## LESSON 5: WILL THE LIGHTS STAY ON? (PART 1)

LESSON SUMMARY

This is the first part of a two-part lesson. In these two lessons children research how electricity is generated in different ways. During this first part children prepare debates about different methods of electricity generation, transmission and the siting of generating plants, which they present to the class in Lesson 6. They learn to recognise which secondary sources are most useful to research their ideas and to use relevant scientific language in their debates.

Preparation required:

Information resources should be placed on the school's intranet site to be accessed by children.

Key vocabulary:

generate, generator, coal, gas, oil, fossil fuels, nuclear, neutrons, atoms, biomass-fired power stations, wind turbine, wave hub, tidal flow, hydro-electric, grid, pylon, transmission, transformer, solar panels

Resources:

Sticky notes, A2 sheets of paper, computers, access to resources to allow children to research electricity generation and using renewable sources of electricity generation

Health and safety:

Re-emphasise the dangers of mains electricity.

National Curriculum links:

There are no direct links to the three statements in the science national curriculum, as these two lessons involve carrying out research and constructing reports about electricity in everyday use

Learning intention:

To research how electricity is generated and transmitted to the classroom, and discuss electricity generation in the future

Scientific enquiry type:

Finding things out using a wide range of secondary sources of information

Working scientifically links:

Reporting and presenting findings from enquires, including conclusions, causal relationships and explanations of and degree of trust in results in oral and written forms such as displays and other presentations

Success criteria:

- I can select information from a range of different sources.
- I can select the best method for presenting the information.
- I can use my scientific knowledge to persuade others in a debate.
- I can use the correct scientific language in a presentation.

This is a research lesson for children to use secondary sources to find out information about different ways of generating electricity. Differing levels of information are provided for children to analyse and present as arguments for and against a range of types of electrical generation in the next lesson. Children work at three levels and the lesson provides an opportunity to move children into different challenge groups after assessing their achievements in the four previous lessons and the Explore part of this lesson.

EXPLORE:

Show children the Making electricity animation (Animation 1) to explain how electricity is made (generated) when a magnet is turned inside a coil of wire. The turning of the magnet can be powered in many different ways.

Ask children, in pairs, to write a sentence on sticky notes summarising what they know about where electricity is generated and how it reaches the home. Display the Making electricity slideshow (Slideshow 1), which gives children words to use: power station, coal, gas, oil, wind farm, pylons, cables, generate, renewable, solar power, water, sun.

Ask children to share these ideas with their partner, using the sticky notes on A2 paper. Ask for an idea from each group and record it on the whiteboard. Ask children to organise their sticky notes into three categories: non-renewable electricity generation, renewable electricity generation and transmission to homes.

It might be necessary to explain what the terms 'non-renewable' and 'renewable' mean when applied to the generation of electricity, if children cannot explain them. This will become apparent as you monitor the groups' discussions about generating electricity. Children's contributions to the discussion will also help you to decide on which challenge they should complete.

Show the Making nuclear electricity animation (Animation 2) on nuclear generation. The nuclear reaction generates heat, which heats water, and the steam turns the turbine to generate electricity.

Key information:

It will be necessary to model one of the methods for generating electricity. Nuclear is more difficult to understand for children and is a good example to choose.

ENQUIRE:

Tell children that they are going to prepare three presentations on: generating and transmitting electricity to the classroom; using renewable electricity generation; and nuclear, wind farm or biomass methods of producing electricity. They are also going to also discuss the subject of siting electricity generating plants locally. Their presentations must not be longer than 10 minutes to allow for questions. Also, provide all three challenge groups with access to the Generating and transmitting electricity video (Video 1), and check that children are accessing relevant information and presenting it in the most appropriate way.

Challenge 1: Children investigate how electricity is generated using coal and gas

Ask the children to investigate how electricity is generated using non-renewable fuels – coal and gas – and how it is transmitted to the classroom. They can use the Making electricity using non-renewable sources (Animation 3) and Transmitting electricity (Resource sheet 1), along with whatever method they think is suitable to present the information, for example, powerpoint or role play. Ask the children to present three key facts about electricity generation and transmission, and to produce a 'Plus, Minus, Interesting' [PMI] on coal- and gas-fired production.

Challenge 2: Children investigate how electricity is generated using renewable sources

Ask the children to research renewable generating sources, such as wind, solar, tidal, wave and hydro-electric, and to prepare a presentation discussing these methods and stating which they think is best for their locality. Ask the children to present two key facts about each method and to produce a PMI with three pieces of information in each section that they consider the most important about their chosen local renewable generating method. They may use the Making electricity using renewable sources animation (Animation 4) and the Renewable sources interactive (Interactive 1).

Challenge 3: Children investigate how electricity is generated using nuclear, biomass and wind sources, and role play for and against arguments for the siting of these electricity generating plants

Ask the children to research nuclear, biomass and wind farm generation of electricity and to present arguments for and against the siting of generating plants in their locality, using role play and class debate. Give them access to the Making electricity using nuclear, biomass and wind resources animation (Animation 5) and the Nuclear, biomass and wind resources interactive (Interactive 2). Explain to the children that the role play is in the form of an interview with the three different plant managers, who will each give three key facts in favour of their plants. The rest of the group has three questions for each plant manager on the disadvantages of each plant. The interviewer then sums up with an interesting fact from each plant, and asks for any questions from the rest of the class.

REFLECT AND REVIEW:

Ask children, in their groups, if they are ready to present their information in the next lesson. Warn them to expect questions from the rest of the class after their presentations. Ask children to write a text message, in their groups, that summarises the most important thing that they have learned. Keep these for the next lesson.

EVIDENCE OF LEARNING:

Listen to children in their groups as they carry out the research. Have they selected the information they require for the debate in their group? Have they chosen the most appropriate method to present the information? Have they shared the work within the group? Have they used scientific language appropriately?

Key information:

Give children figures for the amount of electricity generated by renewable sources (Resource sheet 2)

Key information:

This is a variation on a PMI task. Children undertaking this challenge need to work together to research the information, devise the facts for the plant managers, the questions to ask them, possible answers and the interesting facts that the interviewer will conclude with. They need to decide on the roles to take.

DANGER! LOW VOLTAGE

 ## LESSON 6: WILL THE LIGHTS STAY ON? (PART 2)

LESSON SUMMARY

This is the second part of a two-part lesson. In these two lessons children are researching how electricity is generated in different ways. In the previous lesson children prepared debates about different methods of electricity generation and transmission, and the siting of generating plants. By the end of this lesson they have presented and justified their points of view to the class.

Key vocabulary:

generate, generator, coal, gas, oil, biomass-fired power stations, wind turbine, wave hub, tidal flow, hydro-electric, grid, pylon, transmission, transformer, solar panels

Resources:

Sticky notes, A2 sheets of paper, ICT hardware for presentations as required

National curriculum links:

No direct link to the statements in the Science National Curriculum, as these two lessons are research and report-based on electricity in everyday use

Learning intention:

To present information on how electricity is generated and transmitted to the classroom, and to discuss its generation in the future

Scientific enquiry type:

Finding things out using a wide range of secondary sources of information

Working scientifically links:

Identifying scientific evidence that has been used to support or refute ideas or arguments

Success criteria:

- I can select information from a range of different sources.
- I can decide the best method to present the information.
- I can use my scientific knowledge to persuade others in a debate.
- I can use the correct scientific language in a presentation.

EXPLORE:

Ask each group to share the text message they drafted at the end of the previous lesson. Record these and, at the end of each group debate, ask children if they agree about the importance of the statements made.

ENQUIRE:

Let children know that they have a maximum of 10 minutes for their presentation, after which there will be questions for another 5 minutes or so.

Prepare some questions for each group to answer and use if required.

Challenge 1: Children present information on generating and transmitting electricity

During Lesson 5 the children investigated how electricity is generated using non-renewable fuels – coal and gas – and how it is transmitted to the classroom. In this lesson the children present three key facts about electricity generation and transmission, and a PMI on coal-fired and gas-fired generation.

Ask the children to present their information and answer questions. Then encourage them to discuss the text message from the previous lesson.

Challenge 2: Children present information on generating electricity from renewable sources

During Lesson 5 the children investigated renewable sources for electricity generation, including wind, solar, tidal, wave and hydro-electric, and prepared a presentation discussing these methods and stating which they think is best for their locality. They should have decided on two key facts about each method and generated a PMI with three pieces of information in each section that they consider to be the most important about their chosen local renewable electricity generating method.

Ask the children to present their information and answer questions. Then encourage them to discuss the text message from the previous lesson.

Challenge 3: Children present information on the siting of a generating plant in their locality

During Lesson 5 the children investigated nuclear, biomass and wind farm generation of electricity, and prepared arguments for and against the siting of generating plants in their locality, through role play, to enable class debate.

Ask the children to present their role play, and give them time to engage in debate and answer questions. Then encourage them to discuss the text message from the previous lesson.

REFLECT AND REVIEW:

Ask children whether they think all electricity generation could be based on the use of renewable sources. Children use the figures for renewable electricity generation (Lesson 5, Resource sheet 2) to help them consider whether it is feasible.

Encourage children to discuss the siting of an electricity generating plant locally and to consider which one of the three presented in the role play would be best.

Ask children, in pairs, to record on sticky notes three ways to save electricity, to share their ideas with another pair and to collect a sample of suggestions. Use the sticky notes as part of a display.

Ask: *How would you cope if the lights went out?*

EVIDENCE OF LEARNING:

Do children know about the many methods of generating electricity? Can they select and present the relevant information from secondary sources? Can they use correct scientific language? Can they co-operate to create presentations?

DANGER! LOW VOLTAGE

ENRICHMENT LESSON 1: ARE YOU ALL WIRED UP? (PART 1)

LESSON SUMMARY

This is the first part of a two-part lesson and should be placed after Lesson 4. In this lesson children use their knowledge of electrical circuits and circuit diagrams to construct circuits of different complexity. By the end of the lesson children will have constructed their circuits and started to prepare a presentation about their circuits for Enrichment Lesson 2.

This is an illustrative practical lesson in which children devise their own circuits.

Key vocabulary:

cell, battery, lamp, wire, buzzer, motor, circuit, current, filament, electrical insulator, electrical conductor, mains electricity, switch, terminal, circuit diagram, recognised symbols

Resources:

Cells (1.5V and 4.5V), lamps (1.5V and 2.5V), yellow green red LEDs, flashing lamps (3V), wire, crocodile clips, switches, lamp holders, cell holders, small screwdrivers, wire strippers, wire connection block, corrugated plastic sheet or thick card, paper drill or hole punch, paper fasteners, paper clips, scissors, PVA glue, low-melt glue guns, glue gun stands, digital camera/video camera

Health and safety:

Zinc-carbon and zinc-chloride batteries are the most suitable for this type of activity. Alkaline and rechargeable batteries may become hot enough to cause burns if there is a short circuit. Follow guidance in Be Safe! section 14. Take care with sprung-type wire strippers. Remind children to keep fingers away from the wire cutters incorporated beneath the wire stripper. Low-melt glue guns are cooler than hot-melt guns but still burn. See guidance in Be Safe! section 12.

National curriculum links:

Associate the brightness of a lamp or the volume of a buzzer with the number and voltage of cells used in the circuit, compare and give reasons for variations in how components function, including the brightness of bulbs, the loudness of buzzers and the on/off position of switches, and use recognised symbols when representing a simple circuit in a diagram

Learning intention:

To demonstrate and apply learning about adding and changing components in circuits

Scientific enquiry type:

Carrying out simple comparative and fair tests

Working scientifically links:

Recording data and results of increasing complexity using scientific diagrams and labels, classification keys, tables, scatter graph and/or bar and line graphs

Success criteria:

- I can construct complex circuits.
- I can represent my circuit in a circuit diagram using the recognised symbols.
- I can control components in a circuit with switches.
- I can explain how changing the number and types of components in a circuit affects what happens to them.

EXPLORE:

Show children the Lights video (Video 1) and ask them to think about how the different lights are switched on and off, and how they might be joined in a circuit. Ask them to pick one of the examples shown and to draw a possible circuit using the recognised symbols on mini whiteboards.

Ask children to talk about the circuits in pairs and to select examples to draw on the board.

To keep lamps at the same brightness, another cell must be added in a circuit with two lamps and a switch. Two 1.5V lamps used with a single 1.5V cell are much dimmer because the cell is only producing 1.5 volts, and has 3 volts of lamps to light. Two 1.5V cells in series produce 3 volts, which the two lamps require. Some children might have discovered how to construct parallel circuits in the core lessons and might use this method to keep the lamps the same brightness. Although this is not part of the KS2 programme of study, children should not be discouraged from using parallel circuits, particularly in these challenges.

ENQUIRE:

Explain to children that they are now going to use their knowledge of circuits and circuit diagrams to construct different circuits. Each group should assign roles to different members of the group: design engineers to design the circuit in a circuit diagram and try out the design; build technicians to use the engineers' drawings to construct the circuit and raise any problems with the design engineers; and the public relations (PR) department to record the construction of the circuit with digital images and then present the finished item to the class.

The challenges are differentiated by the difficulty of the task and children should be grouped according to their understanding of the requirements for the different circuits they were shown in the video.

The challenges are presented on the Challenge slides to be displayed on the board, or printed out and placed in the centre of the table. Teacher information regarding the different circuits is provided in the Circuit diagrams sheet (Resource sheet 1).

Key information:

Either coloured LEDs or ordinary lamps can be used. If lamps are used then clip-on lamp holders allow the lamps to be clipped onto lollipop sticks in the correct sequence for traffic lights. LEDs can be mounted on small corrugated plastic sheets with the switch. Switches made in Lesson 2 can be used. It is important that the circuit is a single circuit that operates the three lamps and not three individual circuits that operate independently. Resource sheet 1 shows one suggested circuit diagram and type of switch.

Key information:

It is necessary to explain to children that there are eight lamps to light in the circuit, but not necessarily at the same time. Headlights and tail-lights work together, so all of these must be on at the same time. Indicators work in pairs. With the limitation of the different voltages of cells and lamps, it might not be possible to have the headlights and tail-lights on and the indicators flashing at the same time. In cars the battery and the lamps are all 12 V and can be wired in parallel circuits, giving an equal brightness. Children need to use flashing lamps for the indicators. These are available from suppliers but are often of a higher voltage than other lamps. This higher voltage is required for the lamps to flash.

Challenge 1: Light me to bed – children construct a circuit for an upstairs landing light that can be controlled from two switches that are sited at the top and bottom of stairs

Ask the children to decide who in their group is going to take on which role, and to design and build a circuit. The lamp must light when either of the switches is used. Ask the children to add a second lamp for a downstairs light and ask them whether it is possible to control either lamp from either switch.

Challenge 2: Traffic lights – children construct a single circuit to represent a set of traffic lights that operates in the correct sequence: red, red and amber, green, amber, red

The children present their circuits in the next lesson.

Challenge 3: Wire up my car – children construct a single circuit using cells in series to power head and tail-lights together and direction indicators that flash for left and right, both with switches

Ask the children to mount their circuit on the corrugated plastic (or thick card) with the lights arranged in pairs for the front and back of a car, and the switches in suitable positions. (Slide switch mounted horizontally – move to left for left indicators, right for right indicators.)

The children present their circuit to the class in the next lesson.

REFLECT AND REVIEW:

Group together the design engineers, build technicians and PR departments from the three challenges. Ask each group to write six sentences about the process they went through, including one about modifying what they did, if there was time. Emphasise that they are reporting as scientists. Explain to them that they are going to present their reports in the next lesson.

EVIDENCE OF LEARNING:

Do children co-operate and work as a design team on their challenge projects? Do they demonstrate evidence and understanding of how circuits work in the construction of increasingly complex circuits, including the effects of adding a number of different components, such as extra switches, lamps and cells of different voltages? Are children able to represent their complex circuits in diagrams using the recognised symbols?

DANGER! LOW VOLTAGE

 ## ENRICHMENT LESSON 2: ARE YOU ALL WIRED UP? (PART 2)

LESSON SUMMARY

This is the second part of a two-part lesson. In this lesson children complete their discussions as design engineers, build technicians and PR departments, using information from Enrichment Lesson 1. In their challenge groups children select and present information about their circuits. By the end of the lesson children appreciate the complexity of design required for electrical circuitry in real life and this should be reflected in their presentations.

Key vocabulary:

cell, battery, lamp, wire, buzzer, motor, circuit, current, filament, electrical insulator, electrical conductor, mains electricity, switch, terminal, circuit diagram, recognised symbols

Resources:

Completed circuits from Enrichment Lesson 1, presentation resources as selected by children, for example, projection hardware, recognised symbols cards

Health and safety:

Zinc-carbon and zinc-chloride batteries are the most suitable for this type of activity. Alkaline and rechargeable batteries may become hot enough to cause burns if there is a short circuit. Follow guidance in Be Safe! section 14.

National curriculum links:

Associate the brightness of a lamp or the volume of a buzzer with the number and voltage of cells used in the circuit, compare and give reasons for variations in how components function, including the brightness of bulbs, the loudness of buzzers and the on/off position of switches, and use recognised symbols when representing a simple circuit in a diagram

Learning intention:

To explain and evaluate the design of and construct different circuits and discuss their practicability

Scientific enquiry type:

Carrying out simple comparative and fair tests

Working scientifically links:

Reporting and presenting findings from enquires, including conclusions, causal relationships and explanations of and degree of trust in results in oral and written forms such as displays and other presentations

Success criteria:

- I can select relevant information.
- I can select the best method to present the information.
- I can use my scientific knowledge in a debate.
- I can use the correct scientific language in a presentation.

EXPLORE:

Continuing on from Enrichment Lesson 1, give the design engineers, build technicians and PR departments time to complete their discussions.

ENQUIRE:

Ask children in each group to prepare a presentation about their circuit, its design, construction and any problems. Explain to children that they have a maximum of 5 minutes for their presentation, after which there will be questions for about another 5 minutes.

Prepare some questions for each group to answer and use if required.

Challenge 1: Children present and answer questions about their Light me to bed circuit

During the first part of this two-part lesson the children constructed a circuit for an upstairs landing light, which can be controlled from either of two switches sited at the top and bottom of stairs. The children also added a second lamp for a downstairs light and were asked if it is possible to control either lamp from either switch.

Ask them to present their circuits and answer questions.

Challenge 2: Children present and answer questions about their Traffic lights circuit.

During the first part of this two-part lesson children constructed a single circuit to represent a set of traffic lights that operates in the correct sequence – red, red and amber, green, amber, red.

Ask them to present their circuits and answer questions.

Challenge 3: The children present and answer questions about their Wire up my car circuit.

During the first part of this two-part lesson the children constructed a single circuit using cells in series to power headlights and tail-lights together and direction indicators that flash for left and right, both with switches.

Ask them to present their circuits and answer questions.

REFLECT AND REVIEW:

Ask children, in their challenge groups, to write three true/false statements on mini whiteboards about the circuits that they have constructed. Ask the groups in turn to respond to the other groups' statements.

Ask: *How do you think these challenges are designed and built in the real world? How could you find out?*

EVIDENCE OF LEARNING:

Do children co-operate and work as a design team on their challenge projects? Do they demonstrate evidence and understanding of how circuits work in the construction of increasingly complex circuits, including the effects of adding a number of different components, such as extra switches, lamps and cells? Do children show evidence of being able to overcome problems in design? Do they present the relevant and important information using scientific language?

DANGER! LOW VOLTAGE

 ## ENRICHMENT LESSON 3: CAN YOU PROTECT THE CROWN JEWELS? (PART 1)

LESSON SUMMARY

This is the first part of a two-part lesson. These lessons should be placed after Lesson 4 to use and extend children's knowledge of electrical circuits. In this lesson children use their knowledge of circuits to construct burglar alarms. By the end of the lesson children will have constructed the alarm and begun to design their sales brochure for the alarm, which they will use in Enrichment lesson 4.

Key vocabulary:

cell, battery, lamp, wire, buzzer, motor, circuit, current, electrical insulator, electrical conductor, mains electricity, terminal, switches: pressure, tilt, tremble, slide, toggle, reed

Resources:

Cells (1.5V, 4.5V), lamps (1.5V, 2.5V, 3.5V), flashing lamps (3V), buzzers, wire, crocodile clips, switches from Lesson 2, lamp holders, cell holders, small screwdrivers, wire strippers, wire connection block, corrugated plastic sheet, cardboard box, paper drill or hole punch, paper fasteners, paper clips, metal foil, scissors, PVA glue, low-melt glue guns, glue gun stands, glass beads and/ or plastic tiara, digital camera

Health and safety:

Zinc-carbon and zinc-chloride batteries are the most suitable for this type of activity. Alkaline and rechargeable batteries may become hot enough to cause burns if there is a short circuit. Take care with sprung-type wire strippers. Remind children to keep fingers away from the wire cutters incorporated beneath the wire stripper. Follow guidance in Be Safe! section 14. Low-melt glue guns are cooler than hot-melt guns but still burn. Use glue gun stands. See guidance in Be Safe! section 12.

National curriculum links:

Associate the brightness of a lamp or the volume of a buzzer with the number and voltage of cells used in the circuit, compare and give reasons for variations in how components function, including the brightness of bulbs, the loudness of buzzers and the on/off position of switches, and use recognised symbols when representing a simple circuit in a diagram

Learning intention:

To demonstrate and apply learning about circuits in order to design and construct a burglar alarm

Scientific enquiry type:

Carrying out simple comparative and fair tests

Working scientifically links:

Recording data and results of increasing complexity using scientific diagrams and labels, classification keys, tables, scatter graphs and/or bar and line graphs

Success criteria:

- I can construct a circuit that operates a burglar alarm with audio and visual warnings.
- I can explain how changing the number and types of components in a circuit affects what happens to them.

EXPLORE:

Ask children in groups to make a list of places that require an alarm system. Collate some examples. Ask children what they think would happen if the mains electricity were disconnected to these places and how the alarms would be activated. Remind children of the switches that they made in Lesson 2 and ask how alarms might be activated.

ENQUIRE:

Show the children the Protecting the Crown Jewels video (Video 1). Explain to children that they are going to construct a burglar alarm circuit to protect the Crown Jewels. The alarm will provide three point protection: entry to the Jewel House; entry to the Jewel Room; and moving the Jewels. Other companies are also bidding for the contract. Each of the challenge groups works on one part of the overall alarm system.

Ask children to construct a simple circuit with a lamp, cell and wire. Then ask them to add components – lamps, flashing lamps, buzzers – and ask them what happens. Remind children about matching the voltage of components with the voltage of the cells used.

As in Enrichment Lesson 1, the groups need to appoint design engineers, build technicians and a PR department. The design engineers and the build technicians perform the equivalent tasks as in Enrichment Lesson 1, while the PR department needs to produce a sales brochure for their part of the alarm, which is combined with sections of brochure for the parts of the alarm that other groups are designing. In the second part of the lesson the complete alarm system is put together. The challenges are differentiated by the complexity of the task and children can choose which challenge they wish to work on.

The Circuit diagrams sheet (Resource sheet 1) provides an example of an alarm circuit for each of the three challenges. This can be provided to challenges groups if they are struggling to devise their own circuit.

The challenges are presented on the Challenge slides to be displayed on the board, or printed out and placed in the centre of the table.

Challenge 1: Children construct a Jewel House entrance alarm

Ask the children to construct a circuit that activates when the Jewel House is entered. They must consider whether the alarm is activated by a door, a window or another point of entry. Explain to them that their alarm must activate a flashing warning in the Guard Room and an audible warning around the Tower. It must be able to be turned off by the Yeoman Warders in the Guard Room.

Challenge 2: Children construct a Jewel Room entrance alarm

Ask the children to construct a circuit that activates when the Jewel Room is entered. Explain to them that there are no windows in the Jewel Room, only a door. Explain that the alarm must activate a flashing warning in the Guard Room and a continuous audible alarm around the Tower and in the Warders' barracks. The Yeoman Warders must be able to turn off the alarm in the Guard Room. Remind the children about matching components' voltages with the voltages of cells.

Challenge 3: Children construct a Jewel Box alarm

Ask the children to construct a circuit that activates when the Jewel Box (use a cardboard or corrugated plastic box for this) is opened or moved. Explain that the alarm must activate two flashing warnings in the Guard Room and a continuous audible alarm around the Tower and in the Warders' barracks. The flashing alarms must be able to be turned off before the Warders leave the Guard Room. The Yeoman Warders must also be able to turn off the audible alarm in the Tower and in the barracks.

REFLECT AND REVIEW:

Tell children that although they might not have finished constructing their circuits and sales brochure (they continue this in Enrichment Lesson 4), head office needs to know briefly about any problems the groups have had and how they have dealt with them. Ask children to write an email of no more than 50 words, listing the problems and the solutions. Remind them to report as scientists.

EVIDENCE OF LEARNING:

Do children co-operate and work as a design team on their challenge projects? Do they demonstrate understanding and evidence of how circuits work in the construction of increasingly complex circuits, including the effects of adding a number of different components, such as different types of switches for specific purposes, lamps, cells and buzzers? Are children able to represent their complex circuits in diagrams using the recognised symbols?

Key information:

A door alarm operates when the opening of a door breaks a connection between the door and its frame. This could work using either a tremble switch in the door or a push to break switch in the frame, which is released as the door is opened. The push to break switch is essentially off while the button is pressed, which is when the door is closed. The button is released into its 'on' position as the door is opened and activates the alarm.

Key information:

Children need to match the voltage of components to the voltage and number of cells required to power the series circuit. If children have found out about parallel circuits then components can be wired in parallel and the sound from a buzzer, for example, will be much louder. Circuit diagrams are available for teacher reference.

Key information:

This circuit requires a tilt switch or tremble switch attached to the inside of the lid of the box, which activates when the lid is removed or the box is moved.

DANGER! LOW VOLTAGE

ENRICHMENT LESSON 4: CAN YOU PROTECT THE CROWN JEWELS? (PART 2)

LESSON SUMMARY

This is the second part of a two-part lesson. In this lesson children complete the construction of their burglar alarm circuits and the PR departments present their combined brochure for the burglar alarm to the class. By the end of the lesson children recognise the many applications for which circuits can be used and how, in real life, companies use different departments to fulfil different functions.

Key vocabulary:

cell, battery, lamp, wire, buzzer, motor, circuit, current, electrical insulator, electrical conductor, mains electricity, terminal, switches: tilt, tremble, slide, toggle, reed

Resources:

Completed circuits from Enrichment Lesson 3, presentation resources, for example, projection hardware, as selected by children for the brochure

Health and safety:

Zinc-carbon and zinc-chloride batteries are the most suitable for this type of activity. Alkaline and rechargeable batteries may become hot enough to cause burns if there is a short circuit. Follow guidance in Be Safe! section 14. Take care with sprung-type wire strippers. Remind children to keep fingers away from the wire cutters incorporated beneath the wire stripper. Low-melt glue guns are cooler than hot-melt guns but will still burn. See guidance in Be Safe! section 12.

Key information:

If possible ask the head teacher, another member of staff or a parent to role play the Keeper of the Keys, which will give children the opportunity to make their presentations to an audience and not just to other children in the class. Use Resource sheet 1 to prepare a short brief about the task for the Keeper and some questions to ask.

National curriculum links:

Associate the brightness of a lamp or the volume of a buzzer with the number and voltage of cells used in the circuit, compare and give reasons for variations in how components function, including the brightness of bulbs, the loudness of buzzers and the on/off position of switches, and use recognised symbols when representing a simple circuit in a diagram

Learning intention:

To explain and evaluate the circuits and discuss their practicability as alarms

Scientific enquiry type:

Carrying out simple comparative and fair tests

Working scientifically links:

Reporting and presenting findings from enquires, including conclusions, causal relationships and explanations of and degree of trust in results in oral and written forms such as displays and other presentations

Success criteria:

- I can select the best method to present the information.
- I can use my scientific knowledge in a debate.
- I can use the correct scientific language in a presentation.
- I can construct a circuit with switches that operate a burglar alarm with audio and visual warnings.

EXPLORE:

This lesson is a continuation from Enrichment Lesson 3. Ask children to complete their alarm circuits and their brochure presentations.

ENQUIRE:

Explain to children that the Keeper of the Keys of the Tower wants to see the brochure and hear the presentation about the new alarm system for the Crown Jewels. Each challenge groups' PR department presents its part of the brochure.

Challenge 1: Children present their Jewel House entrance alarm and brochure, and answer questions

In the previous lesson the children constructed a circuit that activates an alarm when the Jewel House is entered by a specified entry point. The alarm needs to satisfy a number of stated criteria.

Ask the children to present their circuit and brochure to the class and to the Keeper of the Keys, and allow time for them to answer questions.

Challenge 2: Children present their Jewel Room entrance alarm and brochure, and answer questions

In the previous lesson the children constructed a circuit that activates when the Jewel Room is entered. The alarm needs to satisfy a number of stated criteria.

Ask the children to present their circuit and brochure to the class and to the Keeper of the Keys, and allow time for them to answer questions.

Challenge 3: Children present their Jewel Box alarm and brochure, and answer questions

In the previous lesson the children constructed a circuit that activates when the Jewel Box is opened or moved. The alarm needs to satisfy a number of stated criteria.

Ask the children to present their circuit and brochure to the class and to the Keeper of the Keys, and allow time for them to answer questions.

Provide the Keeper of the Keys with the Burglar alarm contract (Resource sheet 2), which they can present to the class if they feel the designers, builders and PR departments have demonstrated that their burglar alarm will do the job.

REFLECT AND REVIEW:

Ask children to follow up on the email they wrote to head office in Enrichment Lesson 3, with a second email that contains five bullet points of no more than 25 words each describing the alarm systems. Advise children that they can include a circuit diagram.

EVIDENCE OF LEARNING:

Do children co-operate and work as a design team on their challenge projects? Do they demonstrate evidence and understanding of how circuits work in the construction of increasingly complex circuits, including the effects of adding a number of different components, such as extra switches, lamps and cells? Do children show evidence of being able to overcome problems in design? Do they present the relevant and important information using scientific language?

LIGHT UP YOUR WORLD

INTRODUCTION

In this module children build on the work that they have done in Year 3 where they learned about light sources, how light enables us to see by reflecting from objects and how different objects reflect different amounts of light and shadows. Here they develop a more detailed understanding of mirrors and the reflections that they form, and apply this understanding to make a periscope. They are introduced to ray diagrams that can be used to represent the behaviour of light. They use these diagrams, together with the fact that light travels in straight lines, to explain the formation of shadows and how their size and shape can be affected. They explore refraction in a number of contexts to see how light does not always appear to travel in straight lines. They investigate how white light is made up of many colours of light and how these can be split apart by a prism or in a rainbow, as well as how the colours can be joined together to make white again. In several lessons children engage in illustrative practical activities to explore these phenomena. They also carry out a fair test investigation to measure the size of shadows compared to the relative positions of the light sources, the object making the shadow and the screen.

One key idea in this module is the introduction of ray diagrams as a scientific convention that helps describe the behaviour of light and explains the formation of shadows or images in a pinhole camera. Whilst this may seem relatively straightforward, this move from a descriptive model (the shadow gets bigger, the light reflects from an object to our eyes) to a more scientific model (ray lines can show and predict what is or might happen) can present a significant cognitive jump for some children and you should be considerate of the challenges here. The abstract nature of the ideas represented by the ray diagrams can present a challenge in itself but also represents a shift from a view of light that might be categorised as *sea of light*, where light is all around, to one that draws a single line to represent what the light is doing.

When working scientifically, children ask and propose answers to their own questions about shadow formation as well as exploring quantitatively the formation of shadows. They develop the idea of explaining and supporting the points they make with data and evidence, and consider how confident they feel in the conclusions that they draw, relating them back to predictions that they have made earlier. They carry out illustrative practicals to explore phenomena.

Key vocabulary:

light, dark, shadow, mirror, bright, dim, reflect, eye, opaque, transparent, translucent, ultra violet, ray, beam, refraction, periscope, spectrum, dispersion, inverted, medium, question, investigation, fair test, change, measure, predict, prediction, explanation, observations, draw conclusions

FACT FILE:

The shape of a shadow is defined by the shape of the object causing it. If you move the object creating a shadow towards the light source, the shadow gets bigger. If you move the object towards the screen then the shadow gets smaller. In both cases the shape of the shadow remains the same.

A pinhole camera can work without a lens because the aperture (the hole to let the light in) is small enough to restrict the amount of light let in. The image on the screen is upside down (inverted).

A ray diagram is a model of looking at the behaviour of light that can help predict phenomena such as the size and shape of shadows. The rays are straight lines that travel from the object to the image or the eye, with arrows showing the direction of the light.

When light travels from one medium to another its speed changes. Unless the light travels into the object at right angles to the surface there is also a change in direction. This is called refraction and is why an object half in water appears to be bent and why when we view things underwater they are not where they appear to be.

White light that comes from the sun and other sources, such as a torch, is made up of a number of colours (red, orange, yellow, green, blue, indigo, violet) but we cannot see these because they are mixed together. The light can be split into the separate colours with a prism (dispersion).

White light can be split into its constituent colours in other ways, including using water, and this is how a rainbow is formed. The white light is split by the water drop (rain or mist) but bounces back in the direction it came from, so you can only see a rainbow if the sun is behind you and the rain/mist is in front of you.

Common misconceptions:

Children may think that:

- light is only found in bright areas. If you can see a lit candle from a dark corner of the room, light must be reaching the dark areas of the room for it to have entered your eyes.

- we see things because light travels from our eyes towards an object; the reverse is true - we see because our eyes absorb light rays travelling from the object, reflected from a source.

- objects give out their own light (they actually reflect light from a light source).

- the Moon is a light source (the Moon actually reflects light from the Sun).

- shadows are real 'things' rather than the absence of light (or less light than the surrounding area).

LIGHT UP YOUR WORLD

 ## LESSON 1: WHAT IS LIGHT AND WHAT DOES IT DO?

LESSON SUMMARY:

In this lesson children carry out illustrative practical activities to review their knowledge and understanding about the behaviour of light, including light sources and shadows from Year 3. By the end of the lesson they will be confident to use the key language and ideas needed for this topic.

Key vocabulary:

bright, dark, dim, dull, eye, light, mirror, opaque, reflect, shadow, shiny, translucent, transparent

Resources:

Torches, sunglasses, mirrors, a collection of materials that are transparent, translucent and opaque, a light meter/data logger

Health and safety:

Remind children never to stare directly at any intense light source, such as the sun, a data projector, a laser pointer or a very bright torch. When working in dark areas, ensure floors are uncluttered and be alert to some children's fears of the dark. (see Be Safe! section 5)

Key information:

It is suggested that you aim to try and draw out all of the words in the vocabulary list for this lesson in this activity. Other words that may be useful to introduce here and which will appear later in the module are 'image', 'infrared', 'ultra-violet' and 'nocturnal'.

National curriculum links:

Explain that we see things because light travels from light sources to our eyes or from light sources to objects and then to our eyes

Learning intention:

To consolidate the key ideas from Year 3 about the behaviour of light, including light sources and shadows

Working scientifically links:

Identifying scientific evidence that has been used to support or refute ideas or arguments

Success criteria:

- I can explain that we need a light source, an object and our eyes to see things.
- I can explain why some objects are easier to see than others.
- I can describe what a shadow is and how to make one.
- I know why it is dangerous to look at the sun and how we can protect ourselves.
- I know what the words 'transparent', 'translucent' and 'opaque' mean.

EXPLORE:

Give all children a piece of paper or whiteboard.

Ask: *What are good words to use when talking about light? What do we know about light?*

For each question give them a short period of time to think about their answers on their own and then ask them to write something down. Once they have done this ask them to work in pairs to share their ideas and then in a table group of no more than four children they should try and produce a single sheet with the key ideas on which can then be shared with the class.

You may wish to begin the lesson by showing some of the video assets from Year 3, Module 3, Can You See Me?

ENQUIRE:

Explain to the class that you have a friend who teaches a Year 3 class in another school who needs some help with their lessons. They have a set of questions they want to ask the class in their light topic but they are not sure of the answers to them. Explain to children that the teacher would also like some ideas of things that the class could do to help them answer these questions.

Use Light questions (Resource sheet 1) to show all the questions to the class. Explain to them that they do not need to answer all of them. Stick the individual questions around the room.

How do we see things?

What is darkness?

Why is it easier to find a silver coin than a black button in a dark room?

Is it safe to look at the sun?

What is a shadow?

What do I need to make a good shadow?

What do the words 'transparent', 'translucent' and 'opaque' mean?

Ask the children to work their way around the room and try to answer the questions. Encourage children to present their work in whatever format they wish; it could include images or diagrams as well as writing. You may wish to make available Resource sheet 1, Light questions writing frame, to support them. They can work individually or in pairs.

Explain to children that at the front you have the **ideas box** and if they are stuck they can come and have a look. Inside the box is some of the equipment that is used in the Year 3 lessons to help remind them of what things they might use. (*Some torches, some sunglasses, some mirrors, a collection of materials that are transparent, translucent and opaque, a light meter/data logger, a picture of different light sources.*)

The challenges are differentiated by the level of detail that children need to produce as well as the support provided. Challenge 1 requires them to answer some of the light questions, possibly with the help of a writing frame. Challenge 2 requires them to suggest an activity that children could do to answer one of these questions and Challenge 3 requires them to consider what results other children would get from the activities they suggest.

The challenges are presented on the Challenge slides to be displayed on the board, or printed out and placed in the centre of the table.

Challenge 1: Children answer questions on light

The children try to answer at least four of the questions.

Challenge 2: Children answer questions and provide instructions for an activity for younger children

The children try to answer at least four of the questions and for one of these questions provide some instructions for a light activity that a Year 3 class can do to help them answer the question themselves. They can decide how they want to present the instructions, which may include pictures as well as writing.

Challenge 3: Children answer questions, devise activities and make written predictions

The children try to answer at least four of these questions. For two of the questions they need to come up with a light activity for a Year 3 class to do to help them answer the questions themselves. They write down what they might do and what they might find out.

REFLECT AND REVIEW:

When the Enquire activity has finished, ask children to take their answers to whatever questions they have looked at and leave their answers next to the question. Then give each group of children one question to go and look at. Ask them as a group to collect all the answers to that question, look at all the answers and be ready to share the best ideas with the class. After a short while ask each group to feed back to the class as a whole.

EVIDENCE OF LEARNING:

Look at the answers that they produced in the Enquire activity and listen to children talk and share with their pairs and groups.

Are their answers descriptive and focusing on what happens (for example, the black button is harder to see because it is darker) or do they include some kind of explanation as to why it happens (for example, the black button reflects less light than the silver coin)? Do they use some of the scientific vocabulary correctly?

Can they come up with activities that the Year 3 children could do? Are these activities directly relevant to the question that they are looking at? Are children able to identify the things that they might find out (for example, the darker the object, the harder it would be to see; if the object were opaque it would make a better shadow than a translucent object)?

LIGHT UP YOUR WORLD

LESSON 2: CAN YOU SEE MORE THAN JUST YOUR FACE IN A MIRROR?

Key vocabulary:

light, mirror, reflect, image, reverse, backwards, upside down, inverted

Resources:

Plastic mirrors (one of each for each group), shiny metal spoons (ideally larger than teaspoons so a reflection of a child's face is clearly visible in them)

LESSON SUMMARY:

In this lesson children carry out illustrative practical activities to review and develop their knowledge and understanding of how mirrors work from Year 3, Module 3. By the end of the lesson they are able to describe what happens when different shapes are reflected.

Preparation:

Before the class enters the room place a mirror somewhere discreetly in the class to enable you to see behind you when playing the statues game (see Explore activity).

National curriculum links:

Use the idea that light travels in straight lines to explain that objects are seen because they give out or reflect light into the eye

Learning intention:

To describe how a mirror reflects an image of an object

Scientific enquiry type:

Noticing patterns

Working scientifically links:

Using test results to make predictions to set up further comparative and fair tests

Success criteria:

- I can explain that a mirror works by reflecting light from the surface to my eye.
- I can describe what an object looks like in a mirror.
- I can use evidence from my investigations with mirrors to predict what different shapes and writing will look like when reflected in a mirror.

EXPLORE:

Explain to the class the game 'Statues' (sometimes known as Grandmother's footsteps) where you turn your back and then one of the class start, walking towards you, trying to reach you without you noticing. When you think they have started walking towards you say STOP! Make sure you are standing so that you can see children in the mirror behind you so that you are always be able to win. Once you have won a few times ask one or two of the class in turn to play the game, which they may or may not win.

Explain to the class that you know that you are always going to win because you have cheated and ask them on their own to see if they can work out how you cheated. Share the answers of all the ways you could have cheated and then show them where the mirror is.

Key information:

Try and place the mirror that you are using as discreetly as possible so that children may not see it easily.

ENQUIRE:

Once you have placed the mirror, give children (ideally in pairs) a mirror and ask them to write down all they can in answer to the following questions:

- things that we know about mirrors
- questions we want to ask about mirrors.

These can be collated and referred back to during this and the next lesson.

Display the following question: Does everything look **exactly** the same in a mirror? Explain to children that they are going to try to answer this question by looking at how mirrors work and where they are used, in the rest of the lesson. At this point you should make sure that children are familiar with the term 'image' to describe the 'picture' they see in a mirror and encourage them to use it in their work.

Give each pair one mirror and copies of Mirror activities (Resource sheet 1, 2 or 3 depending on which challenge they are doing.).

The challenges are differentiated by the level of detail that children need to produce as well as their ability to extend their ideas and make predictions. Challenge 1 requires them to carry out a mirror activity and record their observations. Challenge 2 requires them to complete all of the mirror activities as well as considering how to help other children doing this activity. Challenge 3 is an extension of Challenges 1 and 2, and requires children to make predictions about shapes and mirrors. Resource sheet 3 includes some suggestions for Challenge 3 that can be used to support children without reducing the demand, although children should be encouraged to come up with their own suggestions.

Challenge 1: Children complete a mirror activity and record their observations

The children complete the Mirror writing and/or Mirror shapes activities.

Challenge 2: Children complete all the mirror activities and help other children doing the activity

The children complete the Mirror writing and Mirror shapes activities, including recording what they find out about mirrors and shapes, and some hints for other children doing the same activity.

Challenge 3: Children make predictions about shapes and mirrors

Once they have completed Challenge 1 or 2 the children carry out the Predicting the reflection mirror activity, including inventing their own mirror shapes and writing prediction examples.

Once the challenges are completed, return to the children's answers to the earlier statements.

Ask: *What new things about mirrors have you found out? Have you answered any of your questions?*

Encourage the children to use the words identified in the Key vocabulary box, in particular the term 'image'.

Key information:

There are various reverse text generators freely available on the internet as well as within presentation software, so you may wish to prepare some instructions or information to display to the class. Some examples you may wish to use are provided in Resource sheet 4, Reverse Writing Images.

Key information:

In this part of the lesson you may wish to include the word 'inverted' as a scientific term for an image that is upside down.

REFLECT AND REVIEW:

Give children, in pairs, the metal spoons and ask them to look at their reflections and describe what they can see. They should try to explain what is going on and why.

Ask: *How is the spoon making a reflection? What is different about your reflection in the spoon compared with a normal mirror? Why is the image different on the inside of the spoon compared with the outside? Can you describe the whole journey of the light?*

Ask further questions to elicit the understanding that the light is still travelling from the spoon to their eyes. The last question is to elicit the idea that the light goes from LIGHT SOURCE → FACE → SPOON → EYE, just like when children looked in a flat mirror. In this case the curved shape of the spoon is distorting the image. This idea of light travelling from source to object to eye becomes more important in the next lesson.

End the lesson by showing the class the images in Slideshow 1, Reverse writing text, and ask children about what they see in the slides, trying to make connections to what they have done.

EVIDENCE OF LEARNING:

Look at the answers that they produced in the Enquire activity and listen to children talk to each other.

Were they able to see the pattern between the words and shapes and the reflections, noting that the shape is the same, it is just orientation that changes? Did their suggestions and hints in Challenge 2 show that they appreciate the properties of reflection as well as the challenges of carrying out the experiments? In Challenge 3 were they able to make predictions about the shape and orientation of the reflection that were fully or partially correct?

Listen to them as they talk about the reflection in the spoon. In their explanations are they clear about what is happening to the light? Do they describe the role the light source, their face, the spoon and their eyes play in this? Do they correctly order the sequence in which the light travels?

LIGHT UP YOUR WORLD

 LESSON 3: CAN LIGHT GO ROUND CORNERS?

LESSON SUMMARY:

In this lesson children develop their understanding of mirrors from Lesson 2 and use this to develop a model of how light travels. By the end of the lesson they have applied this model practically to make a periscope.

Key vocabulary:

light, mirror, reflect, reverse, backwards, upside down, image, inverted, periscope

Resources:

Plastic mirrors (two for each group), a selection of torches and a small object such as a car or plastic figure (one of each for each group), cardboard, scissors and Sellotape

Health and safety:

Teach children never to stare directly at any intense light source, such as the sun, a data projector, a laser pointer or a very bright torch. When working in dark areas, ensure floors are uncluttered and be alert to some children's fears of the dark (see Be Safe! section 5).

Key information:

Try and use as narrow and bright a beam as possible so that there is a clear beam of light reflected that children can see easily. Experiment in advance to make sure that the reflected beam is clear.

Key information:

Ideally all will be able to make a periscope each; however, this will depend upon the availability of mirrors.

National curriculum links:

Recognise that light appears to travel in straight lines; use the idea that light travels in straight lines to explain that objects are seen because they give out or reflect light into the eye

Learning intention:

To apply understanding of how light travels to explain how a periscope and other applications of mirrors work

Working scientifically links:

Recording data and results of increasing complexity using scientific diagrams and labels, classification keys, tables, scatter graphs, and bar and line graphs

Success criteria:

- I can describe what happens to the light in a periscope to explain how it works.
- I can explain how mirrors can be used to see things that are not directly in line with the eye.
- I can draw a diagram to show how light is reflected from a mirror.

EXPLORE:

Place a large mirror flat on the table in front of you with the reflective side upwards and explain to the class that you are going to turn the lights in the class down very low and shine a bright light onto the mirror. Stand in the position holding the torch (turned off) where you will be and ask them to think, pair, share as to what will happen. Then turn the lights down, turn the torch on and ask children to see if they were right. Turn the torch off and move to a different position and ask them to think, pair, share about what will happen now. At this point remind children that to make a prediction they need to give reasons, otherwise it is just a guess. Ask them to try and use what they saw the first time to help them predict.

ENQUIRE:

Once you have repeated this a few times, explain to children that they are now going to investigate how mirrors work and where they are used. Each challenge has two parts:

1. Round the bend – a series of activities that focus on trying to use mirrors to make light travel around objects. Children only need three mirrors for the third Round the bend activity.
2. Make a periscope.

The challenges are also shown on Resource sheet 1, Round the bend, and Resource sheet 2, Making a periscope. The challenges are differentiated by the level of detail that the children are required to give to describe and explain the activities that they carry out. Challenge 1 is mainly descriptive of the behaviour of the light whereas Challenge 2 requires them to draw diagrams to help describe the way that mirrors reflect light. Challenge 3 asks them to develop these ideas further, adapting the design of a periscope for a different situation.

Arrange children into groups of three or four and give each group three mirrors, a torch and a small plastic figure or similar object, plus scissors, card and sticky tape to make the periscope.

The challenges are presented on the Challenge slides to be displayed on the board, or printed out and placed in the centre of the table.

Challenge 1: Children solve two problems and make a periscope

The children complete the first two Round the bend problems. When they have done this they make their periscope.

Challenge 2: Children solve problems, draw a diagram and make a periscope

The children complete all of the Round the bend problems. When they have done this they collect a blank sheet for at least one of the problems, and try and draw on it how the light is getting from the torch to the eye. Once this is done, they make their periscope and discuss with their group how it works. They need to mention where the light comes from and what happens to it.

Challenge 3: Children solve problems, draw a diagram, make a periscope and discuss how it works

The children complete all the Round the bend problems and draw a diagram showing how the light gets from the torch to the eye. When they have done this they make the periscope and discuss how it works with their group, drawing a diagram to show what happens to the light. Ask the children to make sure they mention where the light comes from and what happens to it. Ask them to try and work out how they could make the periscope differently so they could see behind themselves.

Key information:

Ray diagrams are dealt with in detail in the next lesson and so precise diagrams should not be the focus here. Consideration of the straight line path of the light, the mirror's ability to change its direction and its position, and orientation are the key ideas here.

REFLECT AND REVIEW:

Display the following statements and ask children in groups of three to discuss whether they think they are true or false and how they know:

• light always travels in straight lines

• a flat mirror can make light travel in a curved path

• a periscope works by making light go round corners.

After they have been talking for a while, ask each group to decide which statement they feel the most confident about. Pick one group who are very confident in their understanding for each question and ask them to explain their reasons to the class. Do the others agree?

EVIDENCE OF LEARNING:

Listen to children talk and share with their pairs and groups during the Explore activity. Do they use correct scientific vocabulary? Are their answers descriptive and focusing on what they observe (the beam shines on the ceiling) or do they include some kind of explanation (the light travels from the torch and reflects off the mirror onto the ceiling)?

Listen to them talking and look at the written work during the Round the bend activity. Do they describe what is happening using the LIGHT SOURCE ➜ MIRROR ➜ EYE model of light? Can they represent this in a diagram? Do they make explicit mention of light travelling in straight lines? Are they able to describe how the periscope works? Can they adapt it to look behind them? Do they recognise that the angle and orientation of the mirrors affect where the light is reflected and what can be seen?

Listen to their answers during the Reflect and review activity. Are they able to support their judgments about the statements with examples from the lesson (or elsewhere)?

CROSS-CURRICULAR OPPORTUNITIES:

There is an opportunity to develop the use of mirrors into a technology-based activity where children make a kaleidoscope.

LIGHT UP YOUR WORLD

LESSON 4: CAN YOU MAKE A CAMERA WITH A BOX, PAPER AND A PIN?

Key vocabulary:

light, mirror, reflect, reverse, backwards, inverted, upside down, periscope, ray diagram

Resources:

Plastic mirror, a bright torch, small shoe box, tracing paper, black paper or card (or kitchen foil), scissors and sticky tape, needle or drawing pin; two very large pieces of card, ideally greater than 1 m x 1 m, one with a triangle cut out of it (side length about 50 cm each side) and the other with a smaller circular hole about the size of a tennis ball cut in it; long straight piece of wooden dowel (2 m) or ball of thread, large piece of paper (bigger than A3) and thick marker pen

Health and safety:

Care is needed when using the pin to make a hole for the pinhole camera.

Key information:

A pinhole camera is a simple camera without a lens and with a single small aperture, a pinhole – effectively a light-proof box with a small hole in one side. Light from a scene passes through this single point and projects an inverted image on the opposite side of the box.

LESSON SUMMARY:

In this lesson the idea that light travels in straight lines is reinforced through an illustrative practical activity where children investigate how a pinhole camera works. By the end of the lesson children are able to draw ray diagrams to show how we see objects.

Preparation required:

For the large ray diagram you will need some large (ideally larger than 1m square) pieces of cardboard with holes pre-cut in them. Full details are contained in Resource sheet 3, How to make a big diagram.

National curriculum links:

Recognise that light appears to travel in straight lines; use the idea that light travels in straight lines to explain that objects are seen because they give out or reflect light into the eye; explain that we see things because light travels from light sources to our eyes or from light sources to objects and then to our eyes

Learning intention:

To understand how a pinhole camera works and, using suitable representations, show how this helps us to understand how we see things

Working scientifically links:

Recording data and results of increasing complexity using scientific diagrams and labels, classification keys, tables, scatter graphs, and bar and line graphs

Success criteria:

- I can use the idea that light appears to travel in straight lines to explain how we see things.
- I can draw a diagram showing how the rays of light travel from a light source to an object and then into our eyes.
- I can explain how a pinhole camera works.

EXPLORE:

Place a large mirror on the table in front of you and explain to the class that you are going to turn the lights in the class down very low and shine a bright light down the mirror. Shine the light source so the reflected beam is on the ceiling. Ask children to draw a diagram to show what the light is doing, trying to make it as accurate as possible. Once they have done this, ask them to share their diagram with someone on their table and to agree on the important things that the diagram should include.

Explain to children that diagrams are a scientific way of representing something that happens in the real world (a phenomena), which is why it is important to learn how to draw them properly. Introduce the term 'ray diagram' to describe diagrams which represent light and then show the Slideshow 1, Reflection diagram, as an exemplar here. This includes advice on how to draw good ray diagrams.

Ask the children to look at their diagrams again.

Ask: *Is there anything you wish to change or add?*

ENQUIRE:

Explain to children that they are going to make a pinhole camera, the simplest camera in the world, and explain how it works.

Ask: *What do you think you need to make a camera?*

Draw together their answers, paying particular attention to what the different parts are made of and the path of the light. The list is likely to include lenses but mention that whilst most cameras need lenses, they are going to make a special one called a pinhole camera that can make good images without one. Full details are in Resource sheet 1, Making a pinhole camera.

Key instructions:

Making the camera may be tricky for some children so it is suggested that you have a few spares or part-made cameras available that produce a clear image. You can also use circular crisp tubes to make the pinhole cameras.

The challenges are differentiated by the level of detail that children are required to give about the function of the different parts of the pinhole camera, as well as to link the properties of the materials to their use. Challenge 3 requires children to produce a description and/or diagram of how the light behaves in the camera to make an image.

Challenge 1: Children make a pinhole camera and note its parts

The children make a pinhole camera, noting the important parts of it, what they do and how the light gets from the object to their eyes.

Ask: *What are the different parts of the pinhole camera made of and why? Where does the light on the screen come from?*

Challenge 2: Children make, draw and label a pinhole camera

The children make a pinhole camera. They draw a diagram of the camera, labelling its important parts, what they are made of and why. They should try to use scientific vocabulary.

Ask: *What are the correct scientific terms to describe the properties of the parts of the camera? What would happen if you made the camera from different materials? (such as the screen of black card, the sides of tracing paper).*

Challenge 3: Children make, draw and label a pinhole camera and explain how it works

The children will make a pinhole camera. They draw a diagram of the camera, labelling its important parts, what they are made of and why, using scientific vocabulary. They draw a diagram to show what the light is doing inside the camera and explain the image on the screen as well as considering the full path of the light.

Ask: *Why do you think the image is upside down?*

Key information:

The image on the screen is inverted; this may be unexpected but is because the rays of light from the top of the object go through the hole and to the bottom of the image. Previously children may have said upside down; however, the term 'inverted' should be encouraged, particularly when describing images.

REFLECT AND REVIEW:

Draw together children's ideas from the challenges and then explain to them that you are going to make a giant ray diagram to help explain how the pinhole camera works. Full instructions are in Resource sheet 2, How to make a big diagram (teacher instructions are on Resource sheet 3). Set up the two large pieces of cardboard, screen and torch as shown on the Resource sheet and place the torch inside the triangle, pointing at the screen through the hole.

Ask: *As I move the torch round the inside of the triangle what shape will the bright spot trace out on the screen?*

You may wish to ask a child to make a prediction on the screen beforehand. Then move the torch around the inside of the triangle. This activity is designed to model exactly what is happening with the pinhole camera and show why the image is inverted. You can repeat the activity using a long wooden dowel or piece of thread pulled taut to show the rays of light, thus creating a big ray diagram.

After this activity ask children to draw a diagram of what was happening.

To finish the lesson, talk through with the class the path the light takes all the way from the object you are looking at right to your eyes, extending the discussion beyond just what happens inside the pinhole camera. The LIGHT SOURCE → PINHOLE CAMERA/SCREEN → EYE model of how we see should be emphasised here.

EVIDENCE OF LEARNING:

Listen to children talk as they make, use and talk about their pinhole cameras. Do they use the correct scientific vocabulary for the parts of the camera (opaque for the camera, and black card and translucent for the screen)? Can they describe the image on the screen and what it looks like? Can they use the idea of light travelling in straight lines to explain the image?

Are they able to explain what is happening using the LIGHT SOURCE → PINHOLE CAMERA/ SCREEN → EYE model of light? Can they represent this accurately in a ray diagram? Do they mention the fact that the light travels in straight lines at any point in their explanations or discussions, either unprompted or as a result of non-specific questions such as: What can you tell me about what the light is doing? Do they explain how the diagrams represent the path the light travels?

LIGHT UP YOUR WORLD

 LESSON 5: HOW CAN YOU MEASURE A SHADOW?

LESSON SUMMARY:

This is the first of a two-part lesson. In this lesson children will build on their learning about shadows from Year 3, and about the movement of the Earth in space in Year 5, to plan fair tests to investigate how different variables affect the size of a shadow. They make predictions which they will test in Lesson 6. By the end of this lesson children have identified a variable that they wish to investigate, raised a fair test question and planned how to carry out the investigation.

Key vocabulary:
shadow, opaque, predict, variable

Resources:
Torches, large sheets of white paper, tape measures or metre rulers, card to make shapes, scissors, graph paper

Health and safety:
Teach children never to stare directly at any intense light source, such as the sun, a data projector, a laser pointer or a very bright torch. When working in dark areas, ensure floors are uncluttered and be alert to some children's fears of the dark (see Be Safe! section 5).

National curriculum links:

Use the idea that light travels in straight lines to explain why shadows have the same shape as the objects that cast them

Learning intention:

To identify the variables that affect the size of a shadow, and plan a fair test to investigate one of them

Scientific enquiry type:

Carrying out comparative and fair tests

Working scientifically links:

Planning different types of scientific enquiries to answer questions, including recognising and controlling variables where necessary

Success criteria:

- I can explain how a shadow is formed and what shape it will be.
- I can identify variables to consider when investigating shadow sizes.
- I can decide what measurements to make.
- I can decide what variables to keep the same in my investigation.

EXPLORE:

Show Video 1, Changing shadows. Ask children to think about three questions:

What is staying the same?   What is changing?   Why is it changing?

Once the video has finished, ask children in small groups to share their answers with each other and then draw together some responses from the groups. Establish that:

- the object does not change in size
- the distance between the object and the ground (screen) does not change
- it is the apparent position of the sun relative to the Earth that changes.

ENQUIRE:

Remind children that in Year 3 they observed shadows. Explain to them that they are now going to work in groups of three or four to design a fair test to find out what affects the size of a shadow. This requires them to make accurate measurements. Explain to children that it is important that they are as accurate as they can be, in order to ensure that the measurements are as reliable as possible.

Remind children that fair tests need to be planned carefully, following these steps:

Key information:
You should model the process of generating a question and the development of this into an investigation using an independent variable that it is unlikely or inappropriate for them to select, such as *How does the age of the person holding the torch affect the size of the shadow?*

Step 1   Identify all the variables that could be investigated
Step 2   Select two variables to investigate and raise a question
Step 3   Design an experiment to answer the question
Step 4   Make and justify a prediction

Using a torch and your hand (or another object) make a shadow on the classroom wall, keeping the shadow the same size by keeping them both still.

Ask: *How could I change the size of this shadow?*

Ask children to talk to a partner about the things they could change about how a shadow is formed and to write these on sticky notes. After a short while gather these independent variables together and share them with the class. Ensure that children understand that they are going to test how changing one of these independent variables affects the size of the shadow, the dependent variable.

Explain to them that they are going to carry out the next three steps on their own. Model planning a fair test by raising a question using an independent variable and a dependent variable, in this case the size of a shadow, noting other variables that need to be controlled. Children select their own independent variable to test and then work independently through the next three planning stages.

Ask children to agree their question in their groups and plan a fair test to collect results to answer it. The learning intention is to find out about variables that affect the size of a shadow, so it may be appropriate to steer some children towards a question that will generate valid results, for example, controlling the distance between the screen and the torch, and only changing the position of the object making the shadow.

The challenges are differentiated by the level of detail that children consider in their fair test, the extent to which they justify and explain their decisions, and the detail of scientific ideas applied in their predictions.

The challenges are presented on the Challenge slides to be displayed on the board, or printed out and placed in the centre of the table.

Challenge 1: Children record details of a fair test and make predictions

The children record the details of their fair test. They include an equipment list and detail of results that will be collected and recorded, and indicate which is the independent variable, the dependent variable and which variables will be controlled. They predict the relationship between the independent variable and the shadow size that their results will show.

Challenge 2: Children record details of a fair test, explain their choice of equipment and make predictions

The children record the details of their fair test. They list all the equipment they will need together with what measurements they will take. For some of the equipment they include an explanation of why this is the right equipment to use and/or why other equipment would be unsuitable. They explain how they will make the test fair and how they plan to collect and record their results accurately. They predict the relationship between the independent variable and the shadow size that their results will show, with an explanation of why they think this will be the case.

Challenge 3: Children record details of a fair test, explain their choice of equipment, make predictions and plan how to work as a group

The children record all the details of their investigation including the equipment they plan to use, why this is appropriate, how they will make it a fair test, and how they plan to collect and record accurate results. They predict the relationship between the independent variable and the shadow size that their results will show, with reference to the behaviour of light and the formation of shadows. They also include details about how they plan to work well as a group to complete the investigation.

REFLECT AND REVIEW:

Ask children to place their plans on the tables in the class and then have a short (5 mins) 'marketplace' activity where they are able to walk around the room and look at the work of other groups. They return to their groups and answer the following questions, making changes to their plans if needed.

Ask: *What good ideas did you see in the plans of other groups? After looking at other plans, what do you think are two good things about your plan? What might you change about your plan to make it even better?*

You may wish to collect some of the groups' answers to these questions to share with the whole class, letting children know that next lesson they are going to out the investigation. End the lesson by asking the groups to spend a short while trying to answer the following question.

Ask: *Once we have done the fair test, how can we present our results and share our findings?*

The intention here is to help children to begin to consider what they will do with their results once they have collected them, rather than just focusing on the 'doing part'.

EVIDENCE OF LEARNING:

Read through children's plans, listen to children talking and ask them questions during the planning activity. Can they identify the variables they could change and how to do this systematically? Can they explain their decisions and talk about how this would make a better or fairer investigation? Can they explain why to use a particular piece of measuring equipment and why this is better than an alternative. Do they make decisions about how to work well as a group? Are they able to make some of these decisions and write them down without the need for guidance from the teacher, (such as *Tell me about what things you are doing in this investigation... Why?*) or did they need direct prompting (*Will you keep the distance between the object and the screen the same?*)? Do they mention things that they had done or seen in order to justify their prediction?

In the marketplace activity are children able to identify strengths in the work of other children as well as the strengths in their own work, making improvements as necessary?

LIGHT UP YOUR WORLD

LESSON 6: WHAT DO WE KNOW ABOUT CHANGING SHADOW SIZES?

LESSON SUMMARY:

This is the second of a two-part lesson. In this lesson children carry out the fair test to investigate shadow size that they planned in Lesson 5. By the end of the lesson children have carried out their investigation and presented their findings, and are able to describe the relationship between shape size and the independent variable they selected.

Preparation:

The Big shadow diagram requires pointers, large (>50 cm) pieces of card and screen to be prepared in advance. (See Resource sheet 3 for details.)

Key vocabulary:

shadow, opaque, predict, variable, accurate, reliable

Resources:

Torches, large sheets of white paper, tape measures or rulers, card to make shapes, scissors, graph paper

Health and safety:

Teach children never to stare directly at any intense light source, such as the sun, a data projector, a laser pointer or a very bright torch. When working in dark areas, ensure floors are uncluttered and be alert to some children's fears of the dark (see Be Safe! section 5).

National curriculum links:

Use the idea that light travels in straight lines to explain why shadows have the same shape as the objects that cast them

Learning intention:

To carry out a fair test to investigate the relationship between shadow size and an independent variable

Scientific enquiry type:

Carrying out comparative and fair tests

Working scientifically links:

Recording data and results of increasing complexity using scientific diagrams and labels, classification keys, tables, scatter graphs, and bar and line graphs

Success criteria:

- I can collect and record my results accurately.
- I can present my results in a table or a graph.
- I can use the results from my fair test to describe the relationship between shadow size and an independent variable.

EXPLORE:

Explain to the class that they are going to carry out the fair test they planned in Lesson 4 today. They will be using scientific ideas about light and shadows, and so this first activity is to check what they know already. Arrange them in pairs and give each group a copy of Resource sheet 1, Shadow and light statements, to discuss.

After a few minutes ask each group to pick one statement that they are sure is true, and discuss how they know and what examples they could give to demonstrate this. Remind children that they will need to use some of these ideas when they present the findings from their fair test and that deciding how confident they are in their results is an important part of any scientific investigation.

ENQUIRE:

Children collect their results for the fair test they planned in the previous lesson and then present their findings.

The challenges are differentiated by the way that children present their results, the extent to which they can articulate how confident they are in their conclusion and the evaluation of their own work.

Challenge 1: Children collate results into a table and write an answer to their question

The children collect their results accurately in a table. They can use the blank table template. They write the heading for the table using units as appropriate. They write a conclusion to answer their question in the format: The shadow is biggest when… The shadow is smallest when… They explain whether their prediction was right or not.

Challenge 2: Children collate results in a table and present them in a line graph, and write a conclusion

The children collect their measurements accurately in a headed table, and present the results in a line graph using the line graph template. They decide on the scale, label each axis appropriately and plot the measurements accurately. They write a conclusion to describe the relationship between the shadow size and the independent variable they tested, using two comparatives, such as the greater the distance between the object and the torch, the smaller the shadow. In their conclusion they suggest how confident they are that it is correct as well as whether this is what they predicted. They identify at least one good aspect of their fair test as well as one way to improve it, with an explanation.

Challenge 3: Children collate results and decide how to present them, write a conclusion and suggest how the investigation could be improved

The children collect their results accurately and write up their findings. They decide the best way to present their findings to clearly help answer the question from their investigation, including a graph and possibly diagrams as well. They write a conclusion to describe the relationship they have found. They state whether their prediction was correct or not, how confident they are that this is the case and explain why they think so. They identify how they could adapt or change the investigation so that they could be more confident with their answer.

REFLECT AND REVIEW:

Draw together children's conclusions from the challenges and then explain to them that you are going to look at a giant diagram to try and explain the relationship between the object's distance from the light source and the shadow size that they have found.

Set up the torch, large piece of cardboard, screen and torch as shown in the Resource sheet 2, Shadow size template (teacher instructions are provided on Resource sheet 3). Turn on the torch.

Ask: *What is happening to the light that comes out of the torch?*

Establish that there are lots of light rays coming out of the torch. Remind them of the big diagram you made in Lesson 4 (pinhole camera) and suggest that this time the diagram is going to show them both what is going on and help explain how the shadow size changes.

Attach three pieces of thread to the torch and then ask three children to each take one piece and hold it tight so it just touches the corner of the square. (See the diagram on Resource sheet 2, Big shadow light model.) Give two other children one of the Light here and Shadow here cards and ask them to show where the light and shadow parts on the screen might be. Then move the cardboard triangle towards the screen but ask children holding the threads to stay still for a moment.

Ask: *What happens to the shadow now?*

After children have had a little time to think and talk, ask those children holding the threads to move them back to the corners of the triangle and the children holding the Light here and Shadow here cards to show where the new shadow is. Once this is done, repeat, but this time move the cardboard triangle away from the screen. You may need to offer gets bigger/gets smaller/stays the same size as options for children to vote on.

Following this, you may wish to use Resource sheet 4, Shadow size, to help children consolidate their ideas with the use of diagrams.

EVIDENCE OF LEARNING:

Read through children's results and conclusions, listen to them talking to each other and ask them questions. Are they able to use their results to write a conclusion for their fair test? Do they make reference to whether their prediction was correct or not, and where relevant explain why this was the case? If they have drawn one, does their graph have an appropriate scale and labels, and is the data plotted accurately? Are they able to suggest ways in which their investigation was good and ways in which it could have been better, giving reasons where possible? Are they able to identify the ways in which they worked (or did not work) well as a group? Are children able to say how confident they are that their results are correct?

Key information:

Unlike the version of this activity in Lesson 4, this torch turning on is more symbolic than anything else and the thread traces the light rays.

LIGHT UP YOUR WORLD

LESSON 7: CAN LIGHT CHANGE DIRECTION WITHOUT A MIRROR?

Key vocabulary:

refract, refraction, medium

Resources:

For demo: Glass beaker or clear 1 pt glass, water, cooking oil, pencil, mini whiteboards (if available)

Water magnifier: Clear piece of plastic at least 5 cm × 5 cm with tape around edges, sheet of newspaper, water and dropper if possible, magnifying glass, glass bead

The Surprising Coin: 1p or 5p coin, mug, water

Oil, water and a pencil: A jam jar or similar sized clear, straight edged plastic container mainly filled with water but with a layer of at least a few cm of cooking oil on the top, pencil

Amazing arrows: Jam jar or clear straight edged glass filled with water. Piece of paper with three parallel arrows drawn on it (see Resource sheet 1 and 2, Refraction circus instructions and observation sheets, for more details)

Because of the liquids involved, plenty of paper towels in case of any spillages.

Key information:

At this point you may wish to introduce the term 'medium' as any material that light travels through. Although this term is applied to light in this context, it is used for other waves such as sound, infrared and radio waves.

LESSON SUMMARY:

In this lesson children carry out illustrative practical activities to explore the refraction of light and some of the phenomena it creates. By the end of the lesson they are able to describe refraction as a phenomenon and give some examples of where it occurs.

National curriculum links:

Recognise that light appears to travel in straight lines

Learning intention:

To recognise that whilst light does travel in straight lines, sometimes it changes direction when travelling from one thing into another

Working scientifically links:

Recording data and results of increasing complexity using scientific diagrams and labels, classification keys, tables, scatter graphs, and bar and line graphs; using test results to make predictions to set up further comparative and fair tests

Success criteria:

• I can recognise situations where light does not appear to travel in straight lines
• I can define refraction
• I can describe situations where refraction happens

EXPLORE:

Ask: *Does light always travel in straight lines?* After the last few lessons children will be pretty confident that this is the case. Remind them that this is true but also say that sometimes light seems to do odd things.

Place a pencil in the beaker/glass and ask children to look at it carefully and draw a diagram of what they can see. Then slowly pour water into the glass and ask children to watch very closely as to what is happening. When the glass is almost full, stop and ask them to observe closely and draw another diagram.

Ask: *If you look at all of the pencil does it look completely straight?*

Ask children to talk to a partner about what they have noticed and why it might be happening.

Then explain to the class that light can sometimes do strange things. Explain to them that the scientific word for what is happening is called 'refraction'. Explain to them that there are some experiments set up for them to do today that are going to help them discover more about refraction (and learn some cool tricks to try out at home).

ENQUIRE:

This part of the lesson is based around a circus of experiments that children work through, each with their own instructions and worksheets to support them in their exploration of refraction. There are four different activities but children do not have to do all of them. Set up several versions of each of them around the room so that children can take different amounts of time at each station.

• Water magnifier
• The Surprising Coin
• Oil, water and a pencil
• Amazing arrows

The instructions and worksheets for the activities are included in Resource sheet 1 and 2, Refraction circus instructions and observation sheets. The challenges are differentiated by the level of detail and explanation required from children in their descriptions. Challenge 1 requires children to record their observations, in Challenge 2 children need to add some more description of what is happening and Challenge 3 asks them to add more detail and some attempts to explain what might be happening.

Challenge 1: Children record their observations

The children complete at least two of the activities and write down or draw a picture of what they observed on the sheets provided.

Challenge 2: Children record their observations with explanations

The children complete at least two of the activities and write down or sketch their observations for at least two of them. For at least one of the activities they include some extra information about what they think was happening.

Ask: *What is happening to the light in this activity? What might make the light do that?*

Challenge 3: Children record their observations and add extra details

The children complete and record observations for at least three of the activities. For at least two of the activities they add some extra information about what they think is happening. They must include the oil, water and pencil activity and try to explain, using a diagram, why the oil and water do different things to the light.

Ask: *What is different about the water, oil and air that might affect what the light does? Have you seen any other places where light appears to do the same thing, for example, change direction or appear 'bent'?*

Once the children have completed the activities, draw together some of the ideas and explanations that the children have. You may wish to ask questions such as: What is it that is making the light behave strangely? (the answer being the fact that it goes from air → liquid → air). This happens because the light changes speed when it goes from one medium into another (air → oil, air → water, water → air) and this change in speed causes the change in direction. Whilst this level of detail is not needed to develop an understanding of refraction at this stage, the children may well ask and this is straightforward to explain.

REFLECT AND REVIEW:

Explain to children that knowing about refraction could be more important for some people and animals than they might imagine. Read the first two pages of the extract from Hatchet (Slideshow 1).

Ask: *Is there anything that we have learned in today's lesson that might help Brian?*

Ask them to talk in small groups and decide if they could tell Brian anything that might help him, what would it be? Draw together their answers but don't tell them if they are right or wrong. Explain to the class that he made a bow and arrow, and had the same problem. Then read the second extract from Hatchet (Slideshow 1) and discuss with the class how Brian used his understanding of refraction to help him survive.

EVIDENCE OF LEARNING:

Look at children's work from the refraction circus as well as listening to their conversations and asking them questions. Are they able to describe what they see in a clear fashion? Are their written diagrams and explanations understandable? Do their explanations make reference to both the change in direction of the light and the fact that it changed the medium it was travelling through? Are some of them able to note that whilst both water and oil seem to 'bend' the water, they did so by different amounts?

With the magnifier experiment, did some of children use the magnifying glass or glass bead and try and compare what they saw between their naked eye, the water drop, a magnifying glass and the glass bead?

Listen to their discussion in response to the question about Hatchet. Do they note that the water may make the light bend/refract/change so that things underwater may not be where they appear to be?

Key information:

It is common that the terms 'bent' or 'bending' are used to describe what is happening to the light. Whilst this is not wrong, it may imply that the light is following a curved path, which is not the case. The light follows two straight line paths and just changes direction when it goes from one medium to another.

Key information:

The change in speed of light as it travels through different mediums could be modelled by considering the speed at which a child could run on different surfaces such as a road, mud and water.

Key information:

From the Oil, water and pencil activity, some children might correctly notice that the light seems to 'bend' different amounts between water and air than oil and air.

LIGHT UP YOUR WORLD

 LESSON 8: HOW MANY WAYS CAN YOU MAKE A RAINBOW?

Key vocabulary:

refract, refraction, medium, dispersion, reflect, spectrum

Resources:

Torch (ideally with as white a beam as possible, such as a bike light), a red torch or fairy lights (optional). For each colour wheel: marble (a little bigger than the hole in middle of the CD), glue, CD, white paper, scissors, coloured pencils (red, orange, yellow, green, blue and violet) or printed colour template. (Alternative version without marble will need card/scissors/pens and a strong thread.) Bubble blower and mixture. Tray of water, plastic mirror and torch for each group

Key information:

The dispersion happens when the light passes from the air into the water, just like when it goes from air into the prism. The mirror is needed to reflect the colours out of the water so they can be seen. The light must hit the water at an angle otherwise dispersion does not take place. You may want to ask some children: When you shine white light into a mirror, does it split into colours? to elicit this idea.

LESSON SUMMARY:

In this lesson children carry out illustrative practical activities to investigate how rainbows are made, together with other light and colour effects. By the end of the lesson they know that white light is made of different colours.

National curriculum links:

Recognise that light appears to travel in straight lines

Learning intention:

To understand that white light is made of many colours and these can be separated out

Working scientifically links:

Recording data and results of increasing complexity using scientific diagrams and labels, classification keys, tables, scatter graphs, and bar and line graphs

Success criteria:

- I can make careful observation of situations where white light is split into many colours.
- I can explain how a rainbow is formed.
- I can explain other situations where white light is split into colours.

EXPLORE:

Begin the lesson by shining a torch on the wall.

Ask: *What colour is the light coming out of the torch?*

Hopefully most of them say white (or perhaps yellow or similar depending upon the torch).

Ask: *What colour is the light coming from the sun?*

Explain to the class that we use the term 'white light' to describe light from the sun and torches unless they obviously have a colour. (You may wish to have a red bike light or some fairy lights as well to ask extra questions about.)

Show children Images of rainbows (Slideshow 1) and ask them to think about the question: Where is the light coming from?

Hopefully they note that the light comes from the sun and that can lead on to the question: If the light that you get from the sun is white, where do the colours come from? Inform children that today they are going to look at white and coloured light, how rainbows are made and other places where they might see a spectrum of colours

ENQUIRE:

There are two activities in this section: one shows how coloured light can be combined to make white light and the other shows how white light can be separated out into coloured light. Children do both. Introduce the term 'dispersion' as the scientific word for the splitting of white light into colours and suggest that this might be a useful scientific word in this lesson. Full instructions for the two activities are provided in Resource sheet 1, Making a colour wheel, and Resource sheet 2, Underwater mirror experiment instructions.

The challenges are differentiated by the level of detail, explanation and use of scientific language and diagrams to explain the observations. In Challenge 1 recording is mainly descriptive; Challenge 2 asks children to explain the phenomena in terms of the behaviour of light; and Challenge 3 asks them to try and construct a ray diagram to support their explanation of the spectrum experiment.

Challenge 1: Children carry out activities and record observations

For both of these experiments the children record what they observe in whatever way they think is best. They can use pictures, diagrams, cartoons or text.

Ask: Colour Wheel: *When the wheel spins quickly, can you see the individual colours? Are these colours still there?*

Underwater Mirror: *Where are the colours you see coming from?*

Challenge 2: Children carry out activities and record detailed observations

For these experiments, the children record their observations in as much detail as they can in whatever way they feel is best. They include in their work some information about what might be happening to the light.

Ask: Underwater Mirror: *What happens to the white light when you shine it on the mirror underwater? What scientific terms might you use to describe what is happening?*

Challenge 3: Children carry out activities, record their observations and draw a labelled diagram

The children record all their observations for these experiments. They draw a labelled diagram of the underwater mirror experiment and write down what they think might be happening, using scientific terms and ideas. In their diagrams they include lines to show the rays of light, where they are coming from and where they are going.

Ask: Underwater Mirror: *Can you describe what the light is doing at each step on its journey, from the torch, through the water and mirror, and to your eye? When do you think the light changes and what might be making it change?*

Once the children have carried out both activities and recorded their results, draw together the two activities.

Ask: *Looking at what you did, can you tell me any facts that you now know about white light?* When drawing the ideas together the focus should be on the mixing of colours to make white (colour wheel) and that white light can be split into colours (underwater mirror).

Key information:

Whilst the term 'rainbow' is often used whenever a spread of colours from red to violet is seen, 'rainbow' is a specific term relating to sunlight and water in the atmosphere. The term 'spectrum' **spectrum** is the correct term for the other places where a similar spread of colours is seen.

Key information:

Remind the class of the colours of the spectrum (red, orange, yellow, green, blue, indigo and violet). You could suggest mnemonics such as **R**ichard **O**f **Y**ork **G**ave **B**attle **I**n **V**ain.

Key information:

If it is a sunny day and you have access to a hosepipe then you can create a real rainbow. Turn the hose on and cover the end so you get a fine spray. Stand children with their backs to the sun and spray the water as widely as possible to get fine mist. They should see a rainbow in the spray.

Key information:

It is possible to make a spectrum by shining torchlight onto the playing surface of a compact disk. However, the mechanism for this is more complex as it is caused by many tiny holes on the CD. The underwater mirror activity is designed to show the effect water can have and is related to the formation of a rainbow.

REFLECT AND REVIEW:

Return to Slideshow 1, Images of rainbows, showing them to the class again and ask them to think, pair, share about how they think a rainbow is formed. Show a demonstration of the underwater mirror experiment.

Ask: *What did the water do to the white light?*

Then ask them about when they have seen rainbows and what the weather was like, hoping that they note that this is normally either before, during or after it has rained.

Show them Animation 1, to show the principles of how a rainbow is formed. Make connections between the spectra that they have seen in class with the rainbow and in each case the behaviour of the light. Establish that they have seen a spectrum in class whereas a rainbow is the special name for the spectrum that they have seen in the animation (and real life).

Ask: *Are there any other places where we see spectra?*

As children are talking to each other in small groups, start blowing bubbles into the classroom and ask children to look closely at the bubbles. They should notice that they can see a spectrum in the bubble surface. They may also note that they can see one when oil is spilt on water or by using a prism. Draw these ideas together, noting that in most cases the light needs to go through something else (water, glass, oil, bubble mixture, bouncing off a CD) for it to be split into its colours and air is not able to do this.

EVIDENCE OF LEARNING:.

When talking or writing about the underwater mirror experiment, do they use the terms 'dispersion' and 'white light'? Do they note that the white light is being split up into its colours rather than it being changed by the water? In their diagram are they able to show a single ray of white light splitting into separate colours of the rainbow? You could ask a question such as: When you look at the light from the torch before it goes into the water are all the colours there? to elicit whether children appreciate that the water splits the colours up rather than creates them by changing the water.

When talking or writing about the colour wheel experiment do they note that because the wheel is spinning so fast we see all the colours at once and so the spinner appears white, or at least that it is hard to see the individual colours? You could ask a question such as: When the wheel is spinning, are the individual colours still there?

LIGHT UP YOUR WORLD

LESSON 9: HOW MUCH DO YOU KNOW ABOUT LIGHT?

LESSON SUMMARY:

In this lesson children summarise and consolidate all of the work on light done in Year 6, and assess what they have learned.

Key vocabulary:

shadow, opaque, transparent, translucent, dispersion, medium, inverted, image, refraction

Resources:

Access to the internet or a range of books on light, A3/A2 poster paper and pens

National curriculum links:

Recognise that light appears to travel in straight lines; use the idea that light travels in straight lines to explain that objects are seen because they give out or reflect light into the eye; explain that we see things because light travels from light sources to our eyes or from light sources to objects and then to our eyes; use the idea that light travels in straight lines to explain why shadows have the same shape as the objects that cast them

Learning intention:

To give examples to explain the way that light behaves

Working scientifically links:

Reporting and presenting findings from enquiries, including conclusions, causal relationships and explanations of and degree of trust in results, in oral and written forms such as displays and other presentations

Success criteria:

- I can explain some of the ways that light behaves.
- I can explain how we see things.
- I can use words, diagrams, images and demonstrations to communicate my ideas.

EXPLORE:

Begin the lesson with a quick definition quiz. Randomly pick words from Interactive 1, Light words, and ask children to write down a definition. Then give them an opportunity to think, pair, share and to vote on the best definitions.

ENQUIRE:

Explain to children that you want help to make factual resources for the Year 6 group who will be studying light next year. Explain to them that you want lots of different ideas and ways of presenting what they have learned, and that it should be as interesting and exciting as possible.

Explain to children that they can choose how they present the light ideas; it can be a newspaper article, poster, presentation slideshow, a page from a textbook, a cartoon or anything else that they can think of, as long as the scientific ideas are clearly explained.

Children work individually. Ask them to select one idea about light from Resource sheet 1, Light resources ideas. Explain to them that, as well as the scientific ideas, they should also include at least one new 'light fact' that is something interesting or amazing to do with light. Their audience is other Y6 children who may not know as much as them and so they need to explain any challenging ideas or scientific terms.

The challenges are differentiated by the level of detail and scientific ideas required in the resource as well as the extent to which children are required to consider the needs of the audience. Challenge 3 develops from Challenge 2, asking children to produce a question to test the knowledge or understanding of others.

Challenge 1: Children produce diagrams to explain ideas about light

The children will include in their resource information diagrams that help explain the light idea and make a suggestion for an activity that they could do to help the other children understand the ideas.

Ask: If you could only write one sentence or draw one diagram to explain your light ideas, what would it be? What are the most important words in this resource that the children need to understand?

Key information:

If facilities are available, encourage children to make short videos or use other digital media, although you should monitor the management of time and sharable work output.

Challenge 2: Children produce diagrams and written instructions to explain ideas about light

The children include in their resource information and diagrams that help explain the light idea. They also include details of an activity or experiment that the children could do to help them understand the idea, which should include some diagrams or pictures and instructions to help them.

Ask: *How will your activity help the children learn the ideas about light? What results might they get from the activity? What things might they struggle with and how could you help them?*

Challenge 3: Children produce a quiz or test with a correct answer

Once they have completed Challenge 2, ask the children to write some kind of quiz or test that could be given to the other children to see if they understood the scientific ideas in the resource they have produced, including a correct answer.

Ask: *What makes a really good question? What hints might you give someone doing the test that would help them but not give the answers away?*

Key information:

Some children may struggle to know what to look for when they are to find the 'light fact' to add. Resource sheet 1, Light resources ideas, has some suggestions of areas that children could investigate or research. Examples of good questions from a similar topic would be helpful for children completing Challenge 3.

REFLECT AND REVIEW:

Arrange a 'marketplace' activity where the groups of children go around the room and look at the work of other groups and write some feedback about what they have seen. Resource sheet 2, Light resource review grids, provides a writing frame to support this. It is suggested that each child is given the information resource to review fully (first page on Resource sheet 2) and then the opportunity to look at all of them and record the two best bits of any they see.

EVIDENCE OF LEARNING:

During the Explore activity, note what children say to each other, what they write down, and the feedback and comments that they provide for each other's definitions. Do they define the words correctly?

As children complete their resources, look for the level of detail, use of scientific vocabulary and information they include.

Looking at the work children produce when evaluating other children's work should also provide an insight into their levels of understanding. In particular, their answers to the question: What would you change to make the page even better and why? will be useful here as evaluating the explanations of others can require greater analysis than communicating your own ideas.

LIGHT UP YOUR WORLD

 ## ENRICHMENT LESSON 1: HOW CAN YOU MAKE A GOOD SHADOW PUPPET?

LESSON SUMMARY:

In this lesson children plan, test and make puppets for a shadow theatre. Extension lesson 2 that follows has the performance and evaluation of the activity. Ideally arrange for a class of younger children to be the audience for the performance. It is suggested that these lessons immediately follow Lessons 5 and 6 where children have carried out a shadow investigation, although they could come at the end of the topic or at any point after Lessons 5 and 6.

Key vocabulary:

light, dark, shadow, bright, dim, reflect, eye, opaque, transparent

Resources:

For each of the groups a torch or light source, a piece of tracing or greaseproof paper (ideally A3 in size) in a cardboard frame for stability. These are the practice screens for each group. To make the characters, thick card, scissors and split pins. Thin wooden kebab sticks (with ends cut off and made safe) or similar to attach the figures to

Health and safety:

Teach children never to stare directly at any intense light source, such as the sun, a data projector, a laser pointer or a very bright torch (see Be Safe! section 5).

Key information:

Some children may try to develop complex stories with many characters and not have time to make them. Stories with a few characters are easier to manage.

National curriculum links:

Recognise that light appears to travel in straight lines; use the idea that light travels in straight lines to explain why shadows have the same shape as the objects that cast them

Learning intention:

To design and make shadow puppets and create a design guide that explains how they work

Working scientifically links:

Reporting and presenting findings from enquires, including conclusions, causal relationships and explanations of and degree of trust in results in oral and written forms such as displays and other presentations

Success criteria:

• I can make a shadow theatre.
• I can explain how a shadow theatre works.

EXPLORE:

Begin the lesson by showing the first part of Video 1, Shadow puppet theatre.

Ask: *How do you think the shadow theatre works?*

Get children to share their ideas with each other in small groups (less than five) and then show the second part the video.

Then ask them, in groups of three or four, to write down three or four things they would need to think about if they were making a shadow puppet theatre.

ENQUIRE:

Explain to the class that they are going to work in groups of three or four to make the characters for a shadow puppet story which they will perform in the next lesson. Explain to them that they need to write a short story or adapt one they know and make the figures they need.

Let children know that they are going to present their puppet shows to some of the younger years of the school. They also need to prepare for a question and answer session at the end of the show called How did you do it? where they will be asked questions about how the shadow theatre works. Show children Slideshow 1, the How did you do it? questions, to help them prepare for this. Explain to them that they can prepare their answers in whatever way that they want, including writing, pictures, diagrams or a role-play. Explain to them that at the end of the lesson they are going to share these with each other and give feedback, so they should consider the following criteria:

• how well they explained their ideas
• how scientific their explanations were
• how interesting, entertaining and fun their answers were.

The challenges are differentiated by the detail and complexity of the answers children need to prepare, the level of scientific detail, the use of scientific terminology and diagrams, and how they apply their ideas in a new context.

Challenge 1: Children make a shadow theatre and prepare an answer to a question

The children make their shadow theatre and prepare answers to at least one *How did they do it?* question.

Key information:

Some children may focus on the making of the puppets and not focus on the *How did they do it?* aspect of the task. You could allocate specific periods of time for each, reminding children that they will have very little time next lesson to finish their puppets.

Key information:

Timing may be an issue here and so it is suggested you identify which group answers which *How did they do it?* question.

Challenge 2: Children make a shadow theatre and prepare answers to questions using scientific terms

The children make their shadow theatre and prepare answers to two of the *How did they do it?* questions including key scientific terms and ideas. They also prepare a response to the question: What did we learn from the shadows investigation in the previous lessons that will help us make a better shadow theatre show?

Challenge 3: Children make a shadow theatre and prepare answers to questions using scientific terms and a representation

The children make their shadow theatre and prepare answers to one of the *How did they do it?* questions, including key scientific terms and some kind of diagram or representation (such as a role-play) of what the light is doing. They also prepare a response to the question: What would happen to our puppet show if light did not travel in straight lines, if all opaque materials became transparent and all transparent materials became opaque?

REFLECT AND REVIEW:

Depending upon the group and class size, either perform this task as a whole class or split the class into two or three groups. Ask each group in turn to present their answers to at least one of the *How did they do it?* questions as well as any other questions that their challenges required. Ask each group to consider their scores for the three criteria stated earlier and then select one group to give their feedback. Then ask the group who answered how they might have been able to make their answer even better.

EVIDENCE OF LEARNING:

Listen to the children talking about the shadow theatre throughout the lesson as they make their puppets, and to their responses to questions such as *Why did you pick this material for the puppet? Why is the screen made of tracing paper?* as well as answers to any of the *How did they do it?* questions. Can children use scientific vocabulary in their answer books for scientific terminology (opaque, transparent, translucent)? Can they discuss what the light is doing and the nature of a shadow being the absence of light rather than a 'thing' itself? Their individual presentations of their answers to the *How did they do it?* questions will also provide an insight, as well as the feedback they provide on the answers of others. Are they able to connect the patterns in shadow size and position from previous lessons to the explanations and descriptions they give?

CROSS-CURRICULAR OPPORTUNITIES:

There is significant opportunity for overlap with English/Drama in terms of the script as well as the performance.

LIGHT UP YOUR WORLD

Key vocabulary:

light, dark, shadow, bright, dim, reflect, eye, opaque, transparent

Resources:

The finished shadow puppets from the previous lesson; a large thin white sheet with a large light source behind it for the screen, suspended off the floor so that the children can get behind it to be able to make their shadows without making a shadow themselves

Health and safety:

Teach children never to stare directly at any intense light source, such as the sun, a data projector, a laser pointer or a very bright torch (see Be safe! section 5).

E

ENRICHMENT LESSON 2: WHAT MAKES A GOOD SHADOW PUPPET THEATRE SHOW?

LESSON SUMMARY:

This lesson follows on from Enrichment Lesson 1. In this lesson children perform a story using shadow puppets, applying their knowledge about how puppets are made to create different effects.

National curriculum links:

Recognise that light appears to travel in straight lines; use the idea that light travels in straight lines to explain why shadows have the same shape as the objects that cast them

Learning intention:

To perform a shadow puppet show and explain how it works

Scientific enquiry type:

Noticing patterns

Working scientifically links:

Reporting and presenting findings from enquires, including conclusions, causal relationships and explanations of and degree of trust in results in oral and written forms such as displays and other presentations

Success criteria:

- I can use a shadow puppet to perform a story.
- I can change the way the shadows look.
- I can explain how a shadow theatre works.

EXPLORE:

Begin the lesson with a question.

Ask: *What science ideas about light and shadow did we use when preparing our puppets and show?*

Ask children to write down one idea before they talk to someone else. Give them a minute or two on their own and then share with someone next to them, and then as a group. Draw together these ideas as a class.

ENQUIRE:

Let the class know that they are going to work in the groups from last time and that they have a short period of time to make sure they are ready, and then they are going to perform their puppet shows.

Explain to children that during the performance of each of the groups they need to record some feedback on each group to share. Resource sheet 1, Puppet show review grids, provides a possible frame for this, but you could also use a format such as *two stars and a wish*.

When they have all performed their puppet shows, share some of the feedback with the groups to decide what were the best things about all of the performances and what could have made them even better.

Then inform them that a company has produced a new shadow theatre pack and wants to include some guidance to help someone who wants to perform a shadow puppet show. Children are going to provide the guidance for the pack. Their advice could be a poster, a presentation slideshow, a short video or another format, depending upon resources.

Challenge 1: Children produce information on using shadow puppets

The children produce guidance information that includes information on how shadows are formed, as well as how they can make the puppets appear larger or smaller on the screen.

Challenge 2: Children produce information on shadows and materials for making and using puppets

The children produce guidance information that includes details on how shadows are formed and how they can make puppets appear larger or smaller on the screen. It should also include information about the materials used to make their puppets, using scientific words.

Challenge 3: Children produce information on shadows and materials for making and using puppets, with a diagram

The children produce guidance information that includes details on how shadows are formed and how they can make the puppets appear larger or smaller on the screen. It should include a clear, labelled diagram with information about the materials used to make their puppets, using scientific words. It should include some suggestions about which materials are good and bad to make puppets from and why.

REFLECT AND REVIEW:

Arrange children into small groups of two or three and give them two or three pieces of paper or sticky notes (ideally all the same colour). Ask them to write a question on each one, about light and shadows or anything related to work on the topic so far. Explain to them that they can include questions that they have found the answer to or knew already but, if they can, ask them to also include at least one question to which they are not sure of the answer. Collect these and put them around the room – on the wall or on the tables. Ask children to go round the room and look at the questions, then go back to their groups, write down an answer to that question that they agree on and then place the answer under the question. You could repeat this if wished.

Talk through some of the questions and answers with the group, highlighting any that caused some difficulty or that everyone agreed on. These could be left out and returned to in subsequent lessons, or any question without an answer could be allocated to a child to come back with an answer next lesson.

End the lesson by showing Video 1 from Enrichment Lesson 1 again Human shadow puppet theatre, leading on to a discussion on how they may have created some of the shapes.

EVIDENCE OF LEARNING:

Listen to children as they talk about how the shadow theatre works and look at what they write in their fact sheet. Do they talk about light and shadows, noting where the light, shadow-making object and screen need to be relative to each other? Do they use the terms 'opaque' and 'translucent' for the puppets and screen respectively? Do they make reference to the link between shadow size and the relative positions of the light, puppet and screen, and the investigation that they had previously carried out?

In children's puppet theatre guidance materials, do they present the same ideas? Do they draw diagrams or find other ways to show how the rays of light form the shadow and how the shadow size is affected by its position relative to the screen?

CROSS-CURRICULAR OPPORTUNITIES:

There is significant opportunity for overlap with English/Drama in terms of the script as well as the performance.

GLOSSARY

GLOSSARY

adaptation   the process by which a species becomes better suited to its environment

algae   a division within the Plantae kingdom

amphibian   cold-blooded vertebrate animal, that live in water or on land but must return to the water to reproduce

annelids   a large phylum of segmented worms, with over 17,000 modern species

Animalia   the name of the kingdom that is made up of animals

animal   a living organism

antennae   two long, thin parts attached to the head of an insect and some sea animals. They are used to feel

aorta   the largest artery in your body, it carries oxygen-rich blood pumped out of, or away from the heart

aphid   type of insect

arachnid   living thing, such as a spider, that has a two-part body and eight legs

Aristotle   Greek philosopher, author of works on zoology who contributed greatly to the classification of living things

arteries   blood vessels that carry blood away from the heart

arthropod   animal that has a jointed exoskeleton and jointed limbs

atrium   a chamber in the heart

asthma   a common, inflammatory disease of the airways

backbone   series of vertebrae running from the skull to the pelvis

bacteria   microscopic living things, usually one-celled, that can be found everywhere

birds   warm-blooded animal that has feathers and lays eggs with hard shells

blood   a bodily fluid that delivers necessary substances such as nutrients and oxygen to the cells and transports waste products away

blood vessel   tubes that carry blood around the body

body temperature   a measure of the body's ability to generate and get rid of heat

botany   the study of plants

butterfly   a nectar-feeding insect with two pairs of large, usually brightly coloured wings that are covered with microscopic scales

caffeine   a stimulant found in many drinks

calories   units of measurement used to work out how much energy a particular food supplies

cancer   class of diseases characterized by out-of-control cell growth

capillaries   small (in some cases very small) blood vessels that carry blood through the various tissues of the body

carbon dioxide   type of gas found in the air

carbohydrate   substance that provides energy found in foods such as bread

cell   in a circuit, another word for a battery

cells   basic structural, functional and biological unit of all known living organisms

chrysalis   hard shell spun by a caterpillar, moth or other insect in which a transformation or growth takes place

circulatory system   biological organ system whose primary function is to move substances to and from cells

classification   key way of identifying species or materials through choosing one of two answers to a statement and then moving progressively through statements until an identification is made

clinical trial   research studies that test new ways to prevent, detect, treat or manage diseases

colony   (plural: **colonies**) a group of the same type of plant or animal living together

common characteristics   qualities that living things share

conifers   a division within the Plantae kingdom

conventions   accepted customs

current   a flow of electrical charge carriers, usually electrons or electron-deficient atoms

deoxygenated blood   blood that is poor in oxygen

digestive system   body system that breaks down food that you eat

digestive tract   a part of the digestive system

dispersion   breaking up of light into component colored rays, as by means of a prism

distinguishing characteristics   qualities that set living things apart from one another

division   part of the hierarchy of the plant kingdom, which is as follows: Division > Class > Order > Family > Genus > Species

doping   use of banned performance-enhancing drugs in sports

eatwell plate   highlights the different types of food that make up our diet with recommended proportions

electrical circuit   when components form a complete path through which electricity can flow

electrical conductor   something that does not allow electrical current to flow along it

electrical insulator   something that allows electrical current to flow along it

evolution   gradual process in which something changes into a different and usually better form

extinction   The **extinction** of an animal species occurs when the last individual member of that species dies

fern   one or more of a group of roughly 12,000 species of plants belonging to the botanical group known as Pteridophyta

fibre   a carbohydrate that the body can't absorb, also known as roughage

filament   part of a plant that supports the anther

fish   a vertebrate adapted to living its entire life in the water

fledgling   a young bird that has recently acquired its flight feathers

Fungi   The Kingdom Fungi includes some of the most important organisms

fossil   remains or impression of an animal or plant that lived millions of years ago

generation   an entire body of individuals born and living at about the same time

genes   they carry the information that determines a living thing's traits or characteristics

habitat   a place where an animal finds the things it needs to live. In its habitat an animal finds food, water, and shelter

hatch   cause a young animal to emerge from an egg

heart   organ that pumps blood throughout the blood vessels to various parts of the body

heart rate   is the speed of the heartbeat, specifically the number of heartbeats per unit of time

hibernate   to spend the winter in a deep sleep

inheritance   some variation within a species is inherited which means is passed down from the parents

insects   small living thing that usually has a three-part body, three pairs of legs and two pairs of wings

invertebrates   animal without a backbone

jointed legs   one of the major defining features or characteristics of the arthropods

kingdom   The highest division in the classification system

leaf   (plural: **leaves**) the usually flat part of a plant that makes food for the plant

life cycle   the series of changes occurring in an animal or plant

James Lind   (1716 – 1764) 18th century physician, developed treatments for scurvy

Carl Linnaeus   (1707–1778) famous for his work in Taxonomy, the science of identifying, naming and classifying organisms

lungs   an essential respiration organ

mammal   warm-blooded animal that is covered in hair or fur. Female gives birth to live young and feeds her babies on milk from her own body

metamorphosis   the process of change, for example from a caterpillar to a butterfly

microbes   organism too small to be viewed by the unaided eye, as bacteria, protozoa, and some fungi and algae

micro-organisms   any organism too small to be viewed by the unaided eye, as bacteria, protozoa, and some fungi and algae

migration   the movement of an animal from one habitat or region to another

Phillip Miller   (1694–1771), the head gardener at Chelsea and author of the famous Gardener's Dictionary

mineral   substance in the earth that does not come from animals or plants

molluscs   group of soft-bodied animals such as octopuses, snails and slugs

Monera   the name given to the **kingdom** that contains unicellular organisms that have no nucleus, such as bacteria

mosses   a division within the Plantae kingdom

muscular system   an organ system that helps with movement of the body, maintains posture and circulates blood through the body

natural selection   the way in which a species will gradually evolve

nutrients   substance that is essential for life and growth

nutrition   act of eating and using the goodness in food to grow and stay healthy

opaque   a material that does not let light pass through it

oxygen   gas that is needed by all living things

oxygenated blood   blood rich in oxygen

Plantae   the name of the kingdom that is made up of plants

plasma   a relatively clear, yellow-tinted water containing sugar, fat, protein and salt solution, which carries the red cells, white cells, and platelets

platelets   cell fragments that work with blood clotting chemicals at the site of wounds by sticking to the walls of blood vessels, thereby plugging the gap

press switch   switch that opens or closes an electrical circuit when pressed

protein   substance that the body needs for growth and repair

Protista the name given to a kingdom that is a collection of single-celled organisms that do not fit into any other category

pulse rate your pulse is used to measure your heart rate and can be used to determine how healthy you are

predator an animal that naturally preys on others

prey an animal that is hunted and killed by another for food

pupa an insect in its inactive immature form between larva and adult, for example a chrysalis

push switch switch that opens or closes an electrical circuit when pushed

John Ray (1627–1705) was an English naturalist

RDA recommended daily allowance of a certain food or food group

red blood cell relatively large microscopic cells that normally make up 40-50% of the total blood volume. They transport oxygen from the lungs to the body's living tissues and carry away carbon dioxide

refraction the bending of light as it passes from one substance to another

reptiles cold-blooded animal that has a dry, protective covering of horny scales or plates, and lays eggs with leathery shells

resistor a device used to control current in an electric circuit by providing resistance

rickets condition that affects bone development that is caused by a lack of vitamin D

scurvy disease resulting from a deficiency of vitamin C

selective breeding used to produce new varieties of a species

series circuit an electric circuit connected so that current passes through each circuit element in turn without branching

skeletal system provides the shape and form for a body and protects the internal organs

species part of the hierarchy in the classification system – it forms the second part of an organism's scientific name that identifies one specific organism in the genus

speciation the evolutionary formation of new biological species, usually by the division of a single species into two or more genetically distinct ones

spectrum the group of colours that a ray of light can be separated by

switch device that can be used to open or close the flow of electricity in a circuit

tadpole larva of an amphibian such as a frog, toad, newt, or salamander

terminal device at the end of a wire that allows electricity to flow through a circuit

thorax middle part of an insect's body

toggle switch switch that has a lever that can be moved up or down

translucent a material that lets only some light through

transparent a material that lets light pass through it

variation differences in the same type of plant or animal

vein a blood vessel that carries blood back to the heart. In the veins that collect blood from the body and return it to the heart the blood is deoxygenated

vena cava one of the largest veins in the body

ventricle one of two large chambers in the heart

vertebrate animal that has a backbone

vitamin substance found in the food we eat that the body needs to be healthy

white blood cells cells in the blood that fight germs

Collins are proud to support the work of The Association for Science Education

The Association for Science Education
Promoting Excellence in Science Teaching and Learning

About ASE ...
The ASE is the largest subject association in the UK for teachers of science.
We're a powerful force to promote excellence in science teaching and learning.

Join ASE today ...
ASE membership helps you bring science to life with innovative resources and expert advice that save you time, build your confidence and inspire your pupils.

Why you should join ...
We help you get the basics right! Not every primary teacher has a science background, and teaching science can be a complex process. ASE journals and resources are written by some of the most exciting and experienced science educators in the UK. Our materials help you to master the principles you need to teach, and to understand how young pupils build their knowledge and understanding of science.

We help you excel as a teacher.
ASE membership helps you build your skills as a teacher of science so that you can teach with confidence, generate exciting activities and inspire your pupils to explore and investigate.

Science doesn't have to be difficult – for you or your pupils. We help you to understand science – and how best to teach it.

For more details and how to join, visit www.ase.org.uk/membership/membership-category/primary/

Photo Acknowledgements

p162, top: xpixel/Shutterstock; p162, second: Chantal de Bruijne/ Shutterstock; p162, third: Eric Isselee/Shutterstock; p162, bottom: Eric Isselee/Shutterstock; p163, top: Eric Isselee/Shutterstock; p163, second: Eric Isselee/Shutterstock; p163, third: Erni/Shutterstock; p163, bottom: Eti Ammos/Shutterstock; p164, top: Eric Isselee/Shutterstock; p164, second: Eric Isselee/Shutterstock; p164, third: Eric Isselee/Shutterstock; p164, bottom: Sebastian Knight/Shutterstock; p165, top: Tobb8/Shutterstock; p165, middle: Eric Isselee/Shutterstock; p165, bottom: chris2766/Shutterstock; p175, top left and p179, top left and p183, top left: Chiyacat/Shutterstock; p175, top right and p178, bottom right and p182, bottom left: Elena Pavlovich/Shutterstock; p175, second left and p178, bottom left and p182, bottom right: SF photo/ Shutterstock; p175, second right and p178, third right and p182, third right: Rich Carey/Shutterstock; p175, middle left and p178, third left and p182, third left: Claudia Carlsen/Shutterstock; p175, middle right and p178, second right and p182, second left: TRL/Shutterstock; p175, fourth left and p178, second left and p182, second right: Yuriy Kulik/Shutterstock; p175, fourth right and p178, top right and p182, top right: tomoki1970/Shutterstock; p175, bottom left and p178, top left and p182, top left: tanoochai/Shutterstock; p175, bottom right and p177, bottom right and p181, bottom right: biletskiy/ Shutterstock; p176, top left and p177, bottom left and p181, bottom left: antpkr/Shutterstock; p176, top right and p177, third right and p181, third right: tonanakan/Shutterstock; p176, second left and p177, third left and p181, third left: basel101658/Shutterstock; p176, second right and p177, second right and p181, second right: Olga Anourina/Shutterstock; p176, middle left and p177, second left and p181, second left: smereka/Shutterstock; p176, middle right and p177, top right and p181, top right: Elena Veselova/Shutterstock; p176, fourth left and p177, top left and p181, top left: Spirit of America/Shutterstock; p176, fourth and p179, bottom left and p184, bottom right: Elenamiv/Shutterstock; p176, bottom left and p179, second left and p184, second right: J and S Photography/Shutterstock; p176, bottom right and p179, second right and p184, second left: Stephen Rees/Shutterstock; p179, top right and p183, top right: Francesco de Marco/ Shutterstock; p179, third left and p184, third left: Ulga/Shutterstock; p179, third right and p184, third right: Craig Hanson/Shutterstock; p179, bottom right and p184, bottom right: Oleg Znamenskiy/Shutterstock; p180, top left and p185, top left: Steven Russell Smith Photos/Shutterstock; p180, top right and p185, top right: Creative Travel Projects/Shutterstock; p180, middle left and p185, second left: Daniela Pelazza/Shutterstock; p180, middle right and p185, second right: SNEHIT/Shutterstock; p180, bottom left and p185, third left: rsooll/Shutterstock; p180, bottom right and p185, third right: Patrick Wang/Shutterstock; p183, second left: LehaKoK/ Shutterstock; p183, second right: nopporn/Shutterstock; p183, third left: 06photo/Shutterstock; p183, third right: holbox/Shutterstock; p183, bottom left: Bildagentur Zoonar GmbH/Shutterstock; p183, bottom right: Bernadette Heath/Shutterstock; p184, top left: loflo69; p184, top right: mossolainen nikolai/Shutterstock; p185, bottom left: Tom Grundy/Shutterstock; p185, bottom right: anyaivanova/Shutterstock; p187, top left: Eric Isselee/Shutterstock; p187, top middle: Eric Isselee/Shutterstock; p187, top right: Krasowit/Shutterstock; p187, second left: Steshkin Yevgeniy/Shutterstock; p187, second middle: Eric Isselee/Shutterstock; p187, second right: Eric Isselee/Shutterstock; p187, middle left: Steshkin Yevgeniy/Shutterstock; p187, middle: kazoka/Shutterstock; p187, middle right: ADA photo/Shutterstock; p187, fourth left: Karramba Production/Shutterstock; p187, fourth middle: Shaun Wilkinson/Shutterstock; p187, fourth right: Elena Elisseeva/Shutterstock; p187, bottom left: Eric Isselee/Shutterstock; p187, bottom middle: Henrik Larsson/Shutterstock; p187, bottom right: Eric Isselee/Shutterstock; p188, top left: oksmit/Shutterstock; p188, top middle: Eric Isselee/Shutterstock; p188, top right: Kazakov Maksim/Shutterstock; p188, second left: Eric Isselee/Shutterstock; p188, second middle: Andrey Pavlov/Shutterstock; p188, second right: Eric Isselee/Shutterstock; p188, middle left: cbpix/Shutterstock; p188, middle: Richard Peterson/Shutterstock; p188, middle right: fivespots/Shutterstock; p188, fourth left: QiuJu Song/Shutterstock; p188, fourth middle: panbazil/Shutterstock; p188, fourth right: Cosmin Manci/Shutterstock; p188, bottom left: Dora Zett/Shutterstock; p188, bottom middle: Marek R. Swadzba/Shutterstock; p188, bottom right: Henrik Larsson/Shutterstock; p212: La Gorda/Shutterstock; p213: stihii/Shutterstock; p214: stihii/Shutterstock; p221, top and p223, top and p226, top: designua/Shutterstock; p221, bottom and p223, bottom and p226, bottom: designua/Shutterstock; p222, top and p224, top and p227, top: designua/Shutterstock; p222, top and p224, top and p227, top: amirage/Shutterstock; p247, top: Nattika/ Shutterstock; p247, bottom: Denis Nata/Shutterstock; p248, top: amphaiwan/Shutterstock; p248, bottom: Nuttapong/Shutterstock; p249, top: Valentyn Volkov/Shutterstock; p249, bottom: Melica/ Shutterstock; p250: chrisbrignell/Shutterstock; p277: Ian MacNicol/ Getty Images; p278: Harry Engels/Getty Images; p279: Ian MacNicol/ Getty Images; p281, top left: Bryn Lennon/Getty Images; p281, top second: Dennis Doyle/Getty Images; p281, top third: Michael Steele/ Getty Images; p281, top right: Bryn Lennon/Getty Images; p281, bottom left: Doug Pensinger/Getty Images; p281, bottom middle: Pascal Pavani/Getty Images; p281, bottom right: Pascal Pavani/Getty Images; p282: Clive Mason/Getty Images; p285: ksb/Shutterstock; p287: miqu77/Shutterstock; p313: Dinoton/Shutterstock; p314: holbox/Shutterstock; p315: Natursports/Shutterstock; p372, top: Arena Photo UK/Shutterstock; p372, middle: Nando Machado/Shutterstock; p372, bottom: Lynne Carpenter/Shutterstock